JUST-IN-TIME ACCOUNTING

JUST-IN-TIME ACCOUNTING

HOW TO DECREASE COSTS AND INCREASE EFFICIENCY

STEVEN M. BRAGG

JOHN WILEY & SONS, INC.

New York • Chichester • Brisbane • Toronto • Singapore

Library of Congress Cataloging-in-Publication Data:

Bragg, Steven M.
    Just-in-time accounting : how to decrease costs and
  increase efficiency / Steven M. Bragg.
      p.   cm.
  Includes bibliographical references and index.
  ISBN 0-471-13768-5 (cloth : alk. paper)
    1. Just-in-time systems—Accounting. I. Title.
HF5686.M3B68 1996
657'.068'5—dc20   96-11201

Printed in the United States of America

10 9 8 7 6 5 4 3 2 1

Preface

This book demonstrates for the controller or the chief financial officer how to streamline an accounting system so that transactions can be processed with minimal errors and staff time, resulting in very fast transactions that allow the accounting department more time for business analysis, management reporting, and other activities that contribute to the profits of the company.

Many accounting departments take an excessively long time to process accounting transactions. This problem is caused by many factors, but the following are representative.

Cluttered Systems.   Accounting departments are so overwhelmed by incoming documents and copies of outgoing reports that the burden of processing and filing the information keeps them from spending time on improving processes.

Poorly Documented Systems.   Staff turnover is the bane of an accounting department with no documentation. When an accounting employee leaves or switches positions, the replacement person has no idea of what to do, because there is no information about tasks to perform or sequential steps to complete. As a result, errors caused by new staff people leave errors in the system that must be tracked down and corrected—and that takes time.

Archaic Systems.   Many controllers do not take advantage of new technology or even review their process flowcharts to find easier ways to perform tasks. For example, how many controllers still rely on year-end inventory counts instead of real-time perpetual inventory systems? More advanced techniques like electronic data interchange (EDI) and radio-frequency bar-code scanners are even more rarely used.

Excessive Amounts of Clerical Work.   The proportion of time an accounting department spends on clerical tasks such as reviewing expense reports, entering supplier invoices, generating customer invoices, and applying incoming cash to accounts receivable records can easily exceed 90% of the total labor hours of the department. This leaves little time for analyzing expense patterns, investigating cost reduction projects, advising management about new lines of business or the profitability of existing businesses, or analyzing the department's own inefficiencies.

Excessive Document Wait Times. Many accounting transactions take a startlingly long time to bring to conclusion. For example, an expense report must be prepared by the employee, sent to the accounting department for review, sent to the employee's supervisor for approval, and then sent back to the accounting department for payment. Nearly all accounting transactions, including cash application, time card data entry, and especially accounts payable, suffer from this problem. There are too many transactions in process at one time, which gives the accounting department the impression of being "snowed under" with work.

This book uses a multistep approach to derive methods for improving accounting systems. The first step is to describe and chart an existing process, focusing on the inputs, the processing steps, and the outputs associated with each transaction. Also included, if necessary, is an analysis of wait times and paper movement between employees and departments, in order to differentiate this labor from the labor needed to complete the actual transaction.

Second, a number of suggestions are advanced for reducing the work at various steps:

- Apply new technology to the process.
- Eliminate redundant or unnecessary control points.
- Reduce the number of people involved in processing transactions.
- Remove from the transaction steps that can be handled in advance.
- Automate control points.

The next part of each chapter discusses the effect on accounting controls of the suggestions for improvement. Ways to replace the existing controls with other control points or new controls are discussed. It is very uncommon not to be able to reinforce a deleted control point with controls at another point.

Each chapter also lists the quality issues resulting from the proposed revisions. *Quality* in this book is defined as error-free accounting, so a number of suggestions are made regarding ways to keep errors to a minimum. Each chapter contains samples of cost/benefit analyses that can be used as models when creating actual cost/benefit analyses for justifying the implementation of revised systems. Special problems with identifying costs and associated savings are described.

By streamlining the accounting function, some of the old accounting reports are no longer needed, but new reports are required to control the new systems. These are described in detail in each chapter, and these samples can be used to design actual reports for a new accounting system.

In addition, since the new systems should be measured to compare their performance to that of the systems they are replacing, new performance measurements are provided. Finally, implementation problems that may occur when installing a new system are noted, along with solutions.

The multistep approach just described applies to Chapters 1–7, which concern transactions in the following areas: cash, sales and accounts receivable, inventory, accounts payable, cost accounting, payroll, and the budget.

The second part of the book discusses other accounting areas that require streamlining or are needed in the streamlining effort: electronic data interchange—what it is, how much it costs, how to install it, and its effect on the organization (Chapter 8); and the "quick close"—how to shorten the time required to close the company's accounting ledgers and issue period-end reports (Chapter 9).

One of the first steps in updating an accounting system is documenting the existing system and using that information to identify control weaknesses and error-prone or overlong procedures that must be changed. Chapter 10, Process Documentation, demonstrates how to chart existing and prospective systems, and how to create a procedures manual to train staff in the new systems.

Finally, Chapter 11, Effects of Change on Employees, describes how to avoid problems that arise when substantial changes are implemented. It takes up such issues as stretch goals and their effect on the organization, the need for "hot groups," and retaining key employees while the organization is in the throes of change.

This book is intended for use by the accounting or finance manager who wants to improve the performance of the accounting department. It surveys "best practices" culled from the accounting literature and the author's experience, and describes how to install changes successfully and justify them to upper management.

I would like to thank my editor, Sheck Cho, for the opportunity to publish this book through John Wiley & Sons. Also, my thanks to my wife, Melissa, for tolerating the long hours needed to complete this manuscript.

STEVEN M. BRAGG

Denver, Colorado
July 1996

CONTENTS

JUST-IN-TIME ACCOUNTING

PART ONE

Traditional Accounting Areas

The first half of this book describes ways to increase the speed of operations in several key areas—cash, sales and accounts receivable, inventory, accounts payable, cost accounting, payroll, and the budget. Each chapter details the steps of a typical accounting function and methods for shortening the process by such means as reducing the number of controls, automating procedures, and reducing paperwork transfers among employees.

There are always problems associated with changing systems, and the remainder of each chapter deals with those problems. Topics covered are control issues that arise as a result of specific changes, quality problems, samples of cost/benefit analyses for each suggested change, new reports that are needed to support the changes, technology that can be used to support the changes, ways to measure the altered systems, and possible implementation problems. The intent of these chapters is to provide readers not only with a description of methods for speeding up accounting functions but also with a guide for implementing them.

1

CASH

This chapter reviews a typical cash receipts and disbursements system. It compares the typical system to a modified system that allows a company to process cash inflows and outflows more rapidly. The modifications include the use of lockboxes, overnight investing, zero balance accounts, and modem access to a bank database to review cleared checks. In addition, the type of controls is shown to vary with the *method* of cash receipt and disbursement (e.g., whether cash is received over the counter or through an electronic funds transfer) and the *form* of cash received (e.g., whether it is cash, check, or credit card charge information).

The modified system is then reviewed for control weaknesses and ways to improve the quality of its output. A detailed cost/benefit analysis is presented, which can be used as a model to determine if a company's cash systems should be converted to the modified system. The chapter concludes with a discussion of reports to be expected from the modified system, potential implementation problems, the effect of new technology on the system, and ways to measure its performance.

CURRENT SYSTEM

Cash transactions have been burdened with more controls than any other type of transaction, because cash is easily removed from the company premises and liquidated. This section describes those controls for both cash receipts and disbursements. This information is used later in the chapter to design a cash receipts and disbursements system that involves fewer control points than a traditional system. Fewer controls translate into quicker transaction speed.

The processing of cash receipts for a typical organization generally involves the following steps:

1. All incoming mail not addressed to a specific individual is opened in the mail room.

3

2. Any remittances are logged on a daily remittance sheet prepared in triplicate. The name, check number, date, and amount are recorded.

3. One copy, with the envelopes and remittance slips, is forwarded to the cashier; the second goes to the auditor, treasurer, or controller; and the third is retained in the mail room.

4. The cashier records the cash received via mail on a daily cash sheet, indicating the nature of the receipt along with any other receipts from other sources. This cash record is subsequently sent to the accounting department for posting (details as well as summary), after the cashier has made a summary entry in the records.

5. The deposit slip is prepared in quadruplicate. The cashier retains one copy. Three copies go to the bank for receipting: one is retained by the bank, the second is returned to the cashier as evidence that the bank received the funds, and the third is sent to the auditing department or controller's office.

6. The deposit slips are then compared in total, and occasionally in detail, with the daily cash register. The remittance sheet is also test-checked against the deposit slip. (For control purposes, the cashier does not have access to the accounts receivable records or the general ledger and is not allowed to handle disbursements.)

7. Later, when the monthly statement is received from the bank, an accountant compares the book balance to the bank balance and investigates any variances.

8. The deposit slips are filed by date, and the bank statements are filed by month.

A flowchart for the traditional system of cash receipts (checks only) is shown in Figure 1.1. This traditional cash receipts processing system requires a great many paperwork transfers among employees. The move and wait times thus introduced greatly slow the cash receipts process. Also, every time a piece of paper is transferred, the potential exists for loss or misinterpretation. These two problems—time added and loss of altering of the information—are the same problems encountered on the factory floor when building a product. Since there is a strong similarity, perhaps a review of the existing cash system based on a manufacturing value-added analysis would be appropriate. The value-added analysis shown in Table 1.1 lists each step in the process and the time required to complete each step. One caveat to consider is that there are not really *any* value-added steps in *any* accounting process, because the accounting function does not directly add value to the final product. Thus, in the present analysis, a value-added item is considered to be one that brings the cash transaction closer to conclusion.

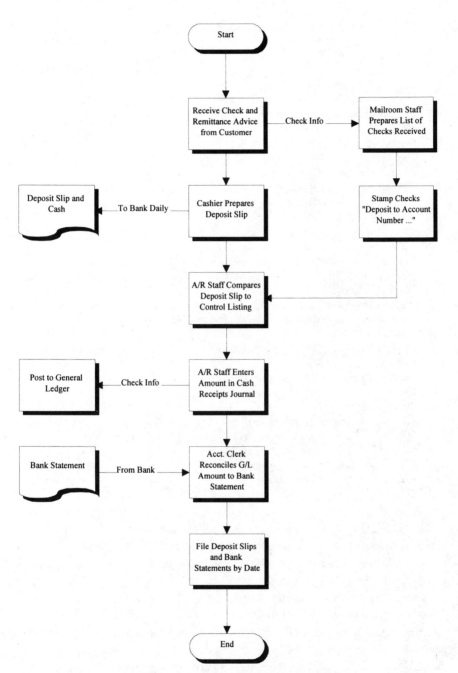

FIGURE 1.1 Cash receipts: Checks only.

TABLE 1.1   Cash Processing Value-Added Analysis

Step	Activity	Time Required (Minutes)	Type of Activity
1	Receive customer cash in mail room.	1	Non-value-added
2	Wait—accumulate until all mail opened.	15	Wait
3	Prepare daily remittance sheet.	5	Non-value-added
4	File copy of remittance sheet.	1	Non-value-added
5	Hand-deliver checks and remittance sheet to cashier.	5	Move
6	Checks wait in cashier's work queue.	60	Wait
7	Cashier prepares deposit slip.	2	Non-value-added
8	Cashier compares deposit slip to remittance sheet.	1	Non-value-added
9	Cashier files copy of deposit slip.	1	Non-value-added
10	Cashier sends copy of deposit slip to accounting mailbox.	1	Move
11	Cashier gives checks and deposit slip to bonded employee.	1	Non-value-added
12	Bonded employee takes checks to bank.	15	Move
13	Bonded employee deposits checks at bank.	5	Value-added
14	Bonded employee brings receipts back to company.	15	Move
15	Bonded employee leaves copies of bank receipts in mailboxes of cashier and accounting department.	1	Non-value-added
16	Wait for employees to check mailboxes.	60	Wait
17	Employees carry receipts to desks.	1	Move
18	Receipts wait in employees' work queues.	60	Wait
19	Employees compare receipt to deposit slip.	1	Non-value-added
20	Employees file receipt with deposit slip.	1	Non-value-added

The table shows optimistic wait times while paperwork waits in employees' mailboxes. In reality, a statistical analysis will reveal a small number of much longer wait times (because of employee absence) that will appreciably affect the completion time of the transaction.

Table 1.2 provides a summary of the value-added analysis. It shows that only 5% of the steps bring the cash transaction closer to conclusion (depositing the cash at the bank); the remaining activities are related to moving paperwork from person to person or making file copies. Several steps exist only to cross-check the information that has been transferred between employees. In terms of time required, the value-added step can be concluded in 5 minutes, while the moving, waiting, and non-value-added steps take up over 4 hours. In short, the action needed to conclude the transaction is only a small part of the total process.

TABLE 1.2 Summary of Cash Processing Value-Added Analysis

Type of Activity	No. of Activities	Percentage Distribution	No. of Hours	Percentage Distribution
Value-added	1	5%	.08	2%
Wait	4	20	3.25	77
Move	5	25	.62	15
Non-value-added	10	50	.25	6
Total	20	100%	4.20	100%

Cash can enter the company in several forms, and by other means than mail. Other forms of cash and methods of receipt are over-the-counter cash sales, credit card sales, and electronic funds transfers (EFTs).

In over-the-counter sales, the primary difference in control as compared with receiving a check in the mail is that the recipient creates a receipt that is given to the customer. This receipt is a substitute for the remittance sheet that was created by the mail room employees for checks received in the mail. Instead of reviewing the control sheet for discrepancies, the internal audit team reviews the sequence of invoice numbers, looking for missing invoices. A flowchart for cash receipts (cash only) is shown in Figure 1.2.

Credit card sales are similar to over-the-counter cash sales in that the recipient must create a receipt and give it to the customer, with the company copy of the receipt being reviewed later for control purposes. The primary difference is that cash is received from the credit card bank in a lump sum rather than as a large number of small cash receipts spread over many transactions. A flowchart for cash receipts (credit cards only) is shown in Figure 1.3.

Electronic funds transfers are quite different from cash, credit card, and check transactions, not because the transactions are electronic but because the cash is received and retained by the company's bank. Since the bank has responsibility for retaining the cash, the number of controls needed for an EFT transaction is appreciably lower than the number used for all other types of cash transaction. For example, control lists of checks received, as well as controls over invoices issued for over-the-counter sales, are not required for EFT transactions. A flowchart for cash receipts (EFTs only) is shown in Figure 1.4.

Besides cash receipts procedures, the controller must also consider the disbursement of cash (Figure 1.5). A typical disbursements system generally involves the following steps:

1. Obtain supervisory approval of a supplier invoice.
2. File invoice by payment date.
3. When due, group invoices for a supplier, and pay for the group of invoices with one check.

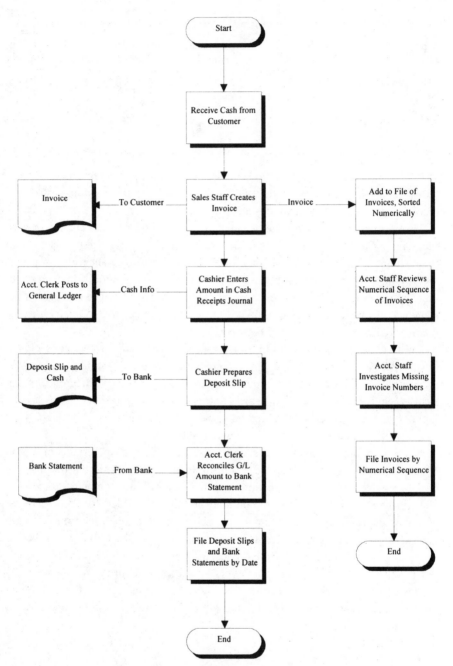

FIGURE 1.2   Cash receipts: Cash only.

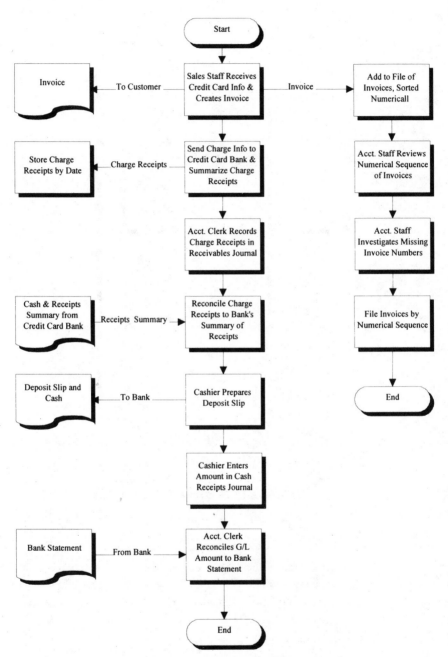

FIGURE 1.3   Cash receipts: Credit cards only.

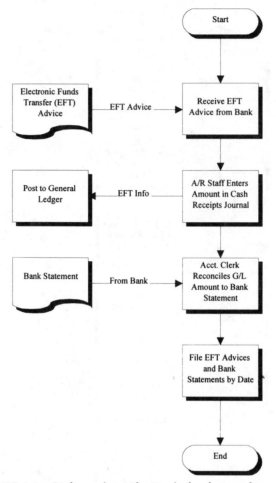

FIGURE 1.4   Cash receipts: Electronic funds transfers only.

4. Record the check in a check register.

5. Forward the check and attached documentation to an authorized check signer, who reviews the supporting documentation and (presumably) signs the check. If there is a problem, the check is returned unsigned, and the accounting staff investigates the problem.

6. Mail the check to the supplier.

7. Stamp the supporting documentation with a PAID stamp and file it.

Most businesses pay for incidental expenses with petty cash funds. The typical petty cash fund operates on an imprest fund basis, so the balance is fixed. At

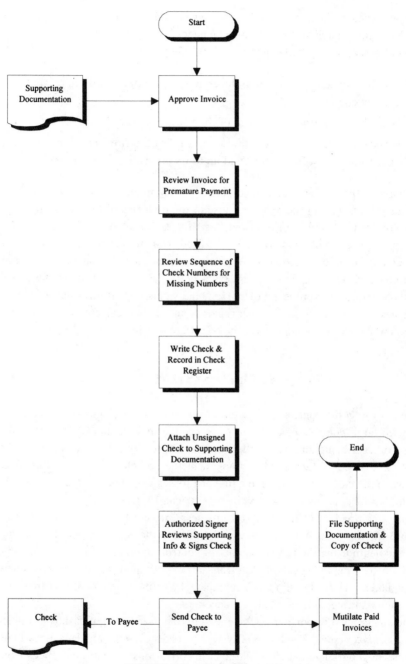

FIGURE 1.5   Cash disbursements.

any time, the cash plus the unreimbursed vouchers should equal the amount of the fund. Numerous funds of this type may be necessary in corporate branch offices. A typical petty cash system generally follows these steps:

1. Obtain a signed voucher from the employee, written in ink (to prevent subsequent changes).

2. Issue cash equaling the amount of the voucher to the employee.

3. When cash amount drops below a certain level, write a check made out to "petty cash" that brings the fund up to its prearranged maximum level.

In summary, the series of actions required to complete a cash receipt or disbursement transaction is complex because of the high risk of theft or inappropriate use of funds. A review of the flowcharts for each type of receipt reveals that the steps required to process an electronic funds transfer are far fewer than those required to process the other types of receipts. In the case of cash disbursements, the number of steps required to process a petty cash transaction are far fewer than those required to process a disbursement by check. This information is used in the next section to design a cash receipts and disbursements system that requires fewer steps to complete than the traditional system.

REVISED SYSTEM

While the foregoing controls may seem burdensome for a company that does not deal with large volumes of cash, they are needed in some industries where the volume of cash being handled is so large that the risk of loss is substantial. The trouble is that controllers tend to implement the complete set of cash controls even when the cash volume is so low that any losses would be minimal. In reality, the number of controls required varies significantly, from many for a casino to few for a company that transacts all business through barter exchange. In this section, we discuss how to break away from using the complete set of cash controls and begin using the fewest number of controls appropriate for the company, thereby increasing the speed of cash transactions. An ancillary discussion is how to speed up the flow of cash into the company by using bank lockboxes, area concentration banking, and zero balance accounts.

The single most important way to speed up the processing of cash is to implement a *lockbox system.* The lockbox system involves establishing depository bank accounts in geographical areas of large cash collections so that remittances from customers will take less time in transit—preferably not more than one day. Customers mail remittances to the company at a locked post office box in the region served by the bank. The bank collects the remittances and deposits the proceeds to the account of the company. Funds in excess of those required to cover costs are periodically transferred to company headquarters. Supporting documents accompanying remittances are mailed by the bank to the company. Collections are thus accelerated through a reduction in transit time, with resultant lower credit exposure. Lockboxes are usually viewed in terms of faster cash deposits into the company's bank account, which improve interest income. However, in terms of faster cash processing, lockboxes allow the company to avoid virtually all controls associated with handling the cash, since responsibility for the cash has now been shifted to the bank. Specifically, the controller can now avoid the following tasks:

- Having the mail room staff create a control list of all checks received.
- Stamping checks "Deposit to Account."
- Preparing a deposit slip.
- Sending the deposit to the bank.
- Comparing the deposit slip to the mail room's control list of checks received.
- Comparing the information recorded by the mail room staff, cashier, and accounting department in search of errors.

The controls used in a lockbox system are shown in Figure 1.6. Notice the reduced number of controls compared with Figure 1.1, which shows the processing of checks received by the company's mail room staff.

Once the controller has set up a lockbox system, the next task is to move as much cash as possible through the lockbox instead of the company's mail room. This is accomplished by contacting all customers and asking them to switch their mailing address for the company to the lockbox address. Including follow-up, this usually takes a number of months.

An extension of the lockbox system is *area concentration banking*. Under this system, company subsidiaries collect remittances (possibly through lockboxes) and deposit them in local banks. From the local banks, usually by wire transfers, expeditious movement of funds is made to a few regional banks. Funds in excess of compensating balances are automatically transferred by wire to the company's banking headquarters. This technique reduces the in-transit time of receivables as well as the mail float involved with mailing checks to corporate headquarters; however, it has no effect on reducing the processing time for cash transactions.

The controller can also maintain a *zero balance account*. With this system, the clearing account is kept at a zero balance. By prior authorization, when checks are presented for payment, the bank transfers funds from the corporate general account to cover the items. Similarly, the treasurer can make wire transfers to the zero bank account on notification of items being presented for payment. This system allows the company to earn interest on funds retained in an interest-bearing

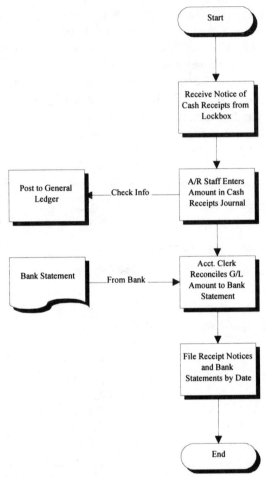

FIGURE 1.6   Revised cash receipts system.

account rather than in a non-interest-bearing checking account; however, it has no effect on the speed of cash transaction processing.

The controller may not be able to take full advantage of the zero balance and cash concentration systems because of compensating balance agreements with banks. These are agreements requiring a company to keep a minimum balance in an account at a bank that has agreed to lend the company money. And drop below this minimum balance incurs penalties. The compensating balance is an indirect way of charging more interest than is shown on a loan agreement, since part of the loan is retained by the bank for its own investment purposes. The employee handling daily cash transactions must factor in the compensating balance before authorizing cash movement between accounts and banks. This is not efficient and leads to the recommendation of *eliminating compensating balance agreements* with all banks. Since a bank uses the compensating balance to earn additional profits, it is likely that the bank will then increase the interest rate on any loans to the company as an alternative source of profits. However, this merely brings the true cost of the loan into the open and eliminates the need to track compensating balances every day.

Several control steps are integrated into the processing of a check used to pay a supplier; some of these steps may be avoided by implementing *electronic data interchange (EDI)* to automatically pay trading partners as services or goods are received. This requires two transactions, one EDI transaction to the trading partner as notification of the transfer and an ACH (Automated Clearing House) transaction to a bank to move the money to the trading partner's account. The EDI transfer allows a company to receive full remittance information and apply it automatically against accounts receivable balances. To ensure that the company can reconcile the EDI transaction containing remittance information to the electronic funds transfer (EFT) that was processed by the bank, the trading partner sending the money includes a reference number in the EFT transaction that references the EDI transaction; this allows the receiving company to match the two transactions when the bank passes along the reference number. By using EDI cash transfers, the controller can now avoid the following steps:

- Approving invoice for payment.
- Reviewing invoice for premature payment.
- Reviewing sequence of check numbers for missing numbers.
- Writing the check.
- Recording check information in a check register.
- Attaching the unsigned check to supporting documentation.
- Signing the check.
- Filing the check and supporting documentation.

Of course, the controller can also use EDI to receive cash from trading partners. When cash is received via EDI, the bank is responsible for the cash and several associated control steps. The following steps can now be avoided by the company:

- Having the mail room staff create a control list of all checks received.
- Stamping checks "Deposit to Account."
- Preparing a deposit slip.
- Sending the deposit to the bank.
- Comparing the deposit slip to the mail room's control list of checks received.

Money can also be moved between trading partners overnight by using a book transfer, the movement of money between accounts at the same bank. If a wire transfer is used, the transfer takes one day. If a paper check is used, the transfer takes one or two days. An EDI remittance advice transaction can transfer money in one day.

In many companies, the controller can safely *cut down the number of approvals* required before a check will be signed. In an extreme (and unfortunately common) case, approval is required on a purchasing requisition document, a purchase order, the supplier invoice, and the check. The controller can require management approval on purchase requisitions and implement strong controls over the use of purchase requisition forms (to prevent authorization forgeries), and skip the remaining approvals (for the purchase order, supplier invoice, and check). An alternative control with the same effect is to require approval of all purchase orders but not of the requisition, supplier invoice, and check. The manual signing of checks is avoided by using a signature plate. There are some exceptions to approving only at the requisition stage, since some items do not have requisitions (such as emergency services and payments for long-term contracts). The controller can even skip all approvals when the dollar amounts are so low that the cost of sending invoices for review is more than the cost of the resulting payment.

Also, since the *petty cash system* has far fewer controls than a disbursement system that uses checks, it behooves the controller to shift as many payments as possible to the petty cash system. These payments are typically the smaller ones that are too expensive in terms of processing effort to pay for with checks. Though petty cash is typically used to reimburse employees, it can be used to pay suppliers on the spot for small amounts; this means the controller will be advocating a voluntary cash-on-delivery system with the company's suppliers, but the total amount of C.O.D. payments will not be large enough to noticeably affect the company's overall cash position.

Finally, the controller can shift disbursements to a *computerized accounting system.* Computerized systems can track the sequence of check numbers automatically (eliminating manual review of check numbers), print only those invoices that are due for payment (eliminating manual review of invoices to deter early payment), and print checks automatically (eliminating manual typing of checks). For those companies that need to print just one check at a time on a fairly regular basis, it may be worthwhile to have a check-printing workstation permanently set up in a locked room, so that single checks can be printed from the computer system. This allows the computer to automatically log new checks into the system, thus avoiding the risk of not recording a check until the bank statement arrives. A flowchart of a revised cash disbursements system is shown in Figure 1.7.

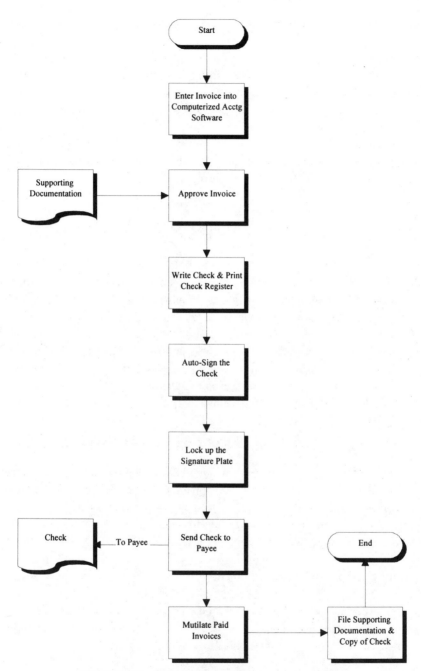

FIGURE 1.7 Revised cash disbursements system.

In summary, the controller can divert incoming checks to a bank lockbox, thereby reducing the number of controls. Similar control reductions can be achieved by receiving and disbursing funds with electronic funds transfers. Also, reducing the number of approvals required to process a check for an expenditure will appreciably speed up the cash disbursement cycle (an analysis of the cost saving is provided later in this chapter). Since fewer controls are involved, disbursing a larger proportion of payments through the petty cash fund instead of through the check disbursement system reduces the overall effort required and increases processing speed. Finally, shifting cash disbursements from a manual to an automated system speeds up the accounting process.

CONTROL ISSUES

There are many control issues related to cash. For example, if the controller were to eliminate controls performed by the mail room employees, the risk of cash payments being intercepted by dishonest employees would increase. Similarly, removing controls over the approval of invoices prior to payment would increase the risk of paying for materials not received. This section contains examples of what would happen if various cash controls were removed in order to increase the speed of cash transactions, and it suggests possible solutions to the resulting control problems. These examples pertain to the changes suggested previously for a revised cash system. Each lists the change and possible revised controls.

Mail Room Stops Recording Checks Received

Having the mail room staff record received checks is a control on the interception of checks. If they are recorded the moment they are received in the mail, the controller has a record that can be verified against the checks later deposited at the bank. If this control is not used, then for example, the cashier could remove checks and attempt to deposit them to a personal bank account.

Move Receipts Away from Mail Room.   The controller can work with the treasurer to have checks go to a lockbox at the bank, thereby eliminating the check interception problem completely. Of course, some work will be required to notify customers of the change of address, so that all checks are mailed to the lockbox. This may take several months to accomplish.

Change the Type of Cash.   Only checks move through the mail room. The controller can work to have a greater proportion of payments sent directly to the bank via wire transfers, thereby avoiding the mail room. Also, payments can be made by credit card, with the check from the credit card company going straight to a lockbox or being wired to the company's bank account. In either case, the risk of cash interception has been eliminated.

Checks Are Not Stamped "Deposit to Account"

This control is used to prevent an employee from cashing a check to a personal account. If the control is removed, the check could be taken out of the group of checks being deposited and put in an employee's account with a similar payee name.

Receive the Cash Elsewhere.   If the cash is deposited in a lockbox, no deposit stamp is needed. Also, if the type of cash is changed from check to credit card or wire transfer, no deposit stamp is needed.

Invoice Reviews Are Not Required

Management approval of invoices is used to verify that payments are being made for services or materials actually authorized and received. If invoice reviews are eliminated, the risk of paying for goods or services that are unauthorized or not received goes up.

Enforce Earlier Approvals.   Most companies require management approval of purchase requisitions or purchase orders, so earlier approvals may already be included in the system. As long as all purchases are authorized with a signed purchase order, the risk associated with invoice approval is reduced. However, there are usually a moderate number of invoices that have no associated purchase requisition or purchase order—these are for long-term contracts at a set fee (such as lawn care) or emergency service (such as computer repairs) for which it would have been detrimental to take the time to procure a purchase order. For invoices for emergency service, the invoice approval should be retained, since it is the only control point in the process. Invoices for long-term contracts should also be reviewed regularly, because they are subject to a particular hazard—the payables staff can become so used to paying the same amount each month that they neglect to note the termination date of the contract and continue to pay past that date.

Use Dollar Limits.   Many invoices are for negligible amounts. The controller can set a dollar limit below which all invoices are automatically approved for payment. Invoices for smaller amounts are not worth the review cost. In case many small invoices were being used to circumvent the dollar limit approval level, the controller could review the total number of payments made to suppliers to see if a disproportionate amount were being paid to individual suppliers without management authorization.

Checks Are Not Reviewed for Premature Payment

This control is used to keep the payables department from paying invoices prior to their due dates. This is a minor control, since the only loss is the investment value of the money during the period when it would normally have been retained. A nearly identical control is reviewing invoices to ensure that early payment is being made on invoices that allow early payment discounts.

Use Computerized Date Tracking.   Most computerized accounting systems have a feature that lists all invoices needing to be paid early in order to claim discounts as well as invoices that are currently due for payment—all other invoices in the system are excluded from the report until they become due for payment.

Sequence of Check Numbers Is Not Reviewed

This control is used to verify that no checks have been removed from the check stock and improperly issued. If the review is no longer conducted, a check could be removed from the check stock with no risk of detection and written to an unauthorized payee.

Use Computerized Sequencing of Numbers.   The check numbers can be stored in a register in a computerized accounting program, so that the next number is automatically assigned to a blank check by the computer. Of course, the check stock must still be safeguarded so that unnumbered checks are not stolen. Also, access to the computer software voids the control (since anyone can then print a check), so access to the software must be guarded, either physically or with passwords.

Complete the Bank Reconciliation.   A carefully completed bank reconciliation will reveal the presence of checks outside of the normal sequence of check numbers. However, this control is after the fact; the check will have already been cashed, so the controller can only investigate and hope to collect the money from the offending party.

Checks Are Not Manually Signed

This is the last control before a check is mailed to a payee, since the check signer is assumed to be reviewing the checks and any accompanying documentation. If checks are not manually signed (being printed by a computer printer or an imprinting device), it is possible for checks to be improperly issued to the wrong parties.

Enforce Earlier Approvals.   Management approval of purchase requisitions or purchase orders (and sometimes invoices as well) makes the review by the check signer redundant.

Batch Payables.   If early approvals are used as a control solution, then only invoices with an approval stamp should be paid. To prevent checks from being issued for unapproved invoices, all approved invoices should be processed in batches with a control sheet that summarizes the approved invoices in the batch. If the check register varies from the total on the batch summary, the variance should be investigated to ensure that an unapproved check has not been inserted into the batch.

Petty Cash Is Used to Pay for More Transactions

Petty cash is less controllable than payment with checks because it tends to fall outside tight controls like purchase requisitions and purchase orders. More use of petty cash can lead to large volumes of cash being paid out in small increments.

Obtain Supervisor Approval.   If supervisory approval is required before petty cash is disbursed, the risk of inappropriate disbursement is reduced.

Obtain Audit Review.   The internal audit department should review the types of disbursements being made, especially those that, on a cumulative basis, add up to a large sum. These expenditures may indicate that cash is being spent on items that should be subject to greater review through the check authorization process.

In short, none of the preceding control changes involve simply removing a control, because the risk of loss would increase dramatically. Instead, they call for shifting the type of cash to one that requires fewer controls, shifting to a different method of receipt or disbursement, or moving controls to a different point in the process.

These changes are minimal when compared to the long list of cash controls that are still necessary in nearly all corporate environments. The following controls should be maintained to avoid the potential increase in risk associated with their elimination.

Divide Responsibility

The controller can assign one person to handle all cash-related transactions but then runs the risk of cash losses. For example, if one person writes checks, signs them, and performs the bank reconciliation, that person could write a check to a personal account and hide all traces of the transaction. Thus, the responsibility for cash transactions should be divided among employees on the general principle that whoever handles cash should not be allowed to record cash transactions. For example, whoever opens the mail should not be the person who applies payments against receivables records. In addition, the person who approves payables for payment should not be the person who creates checks. Finally, the person who performs the bank reconciliation should not sign checks or approve payments.

Shift Jobs Frequently

Nearly all the controls mentioned in this chapter can be circumvented by the collusion of two or more employees. The best way to prevent collusion is to continually shift job responsibilities among employees, so that collusions will, at best, be short-lived.

Enforce the Taking of Vacations

Often, embezzlement can only be continued if the perpetrator is on site at all times. For example, if a payroll clerk keeps nonexistent employees on the payroll and then pockets the payroll checks, a vacation during the next pay period will reveal the error, for some checks will be left over and unpocketed.

Use Periodic Audits

Many controls do not have to be performed continually. Instead, they are conducted on a "spot" basis; these are ideal tasks for the internal audit group. The following items, as well as many of the other controls listed in this section, can be assigned to the internal auditors.

Review Supplier Relations.   The company's internal audit group should periodically review contracts with long-term suppliers to verify the amount of payments made, as well as their bona fide existence, and any relations with or ownership by company employees.

Compare Cash Register Totals to Cash Counts.   It is possible that cash receipts are not being logged in via the cash register. To detect this problem, the cash count should be compared to the register total and cash register procedures should be reviewed.

Verify Cash Sales against Inventory Records. Except for service industries, when a cash sale occurs, inventory is given to the customer in exchange for cash. Therefore, the internal audit group can compare inventory issuance amounts and dates to the amounts and dates of cash receipts to uncover variances.

Compare Mail Room Control Totals to Deposit Slips. This comparison reveals any discrepancies between the total amount of checks received and the amount deposited. Variances indicate that checks are being removed prior to being deposited at the bank.

Compare Payroll Reports to Actual Headcount. A payroll clerk can easily hand out checks to the staff and cash any that are left over. This ploy can continue for some time if the same clerk has control over when someone is deleted from the company payroll. The internal auditors can match workers to paychecks, and look for leftover checks. If payroll checks are mailed directly to employees or are direct-deposited, the auditors can look for multiple checks going to the same address or bank account number.

Trace Cash Receipts to Related Receivables. A receivables clerk who also has control over cash received can divert funds by crediting funds received from one customer against another account from which cash has been diverted earlier (called "lapping"). Tracing specific cash receipts to customer receivables balances (and paying particular attention to the dates on which accounts were credited) will spot this control problem.

Review Value of Trade-Ins. Some companies accept trade-ins of old equipment when new equipment is purchased by customers. In these situations, salespeople can inflate the value of trade-ins to the company and pocket the cash difference between the price paid and the value of the trade-in as represented to the customer. To control this problem, the internal audit group can compare the recorded value of trade-ins to independent appraisals of their values. Another control measure is to verify the amount of the cash sale with the customer.

Deposit All Checks Daily

If checks are only sent to the bank infrequently, the checks that are stored on site until the next deposit are subject to theft. This is also bad cash management, since the funds would otherwise be earning interest. The proper control is to deposit checks every day or to avoid the procedure entirely by setting up a lockbox.

Always Issue Receipts

Require all employees who receive cash (e.g., salespeople, tellers, store clerks) to issue receipts to customers for the cash received. A duplicate is always retained, so that the internal audit staff can reconcile the amount of cash received to the amount listed on the receipts. A variance may indicate that cash was withheld by the employee. This can be a weak control, since the employee may neither create a receipt nor turn in the collected money. Among the most difficult forms of theft to detect is one where no receipt is issued in exchange for a check made out to "cash." In this case, the bank will always cash the check and pay the employee; if the check had been made out to the company, the bank might have notified the company of potential wrongdoing.

Even when receipts are issued to customers, the salesperson may later write in a discount, which is then withheld from the company. The best controls for this are reviews of receipt copies for evidence of tampering and follow-up calls with customers to compare company sales records to their receipt records.

Closely Review Accounts Converted to Bad Debts

The controller should thoroughly review requests to write off bad debts. Though this may appear to be a control over accounts receivable, it is also a control over cash; an employee can write off an account as a bad debt and then pocket the check when it arrives. When reviewing the bad debt request, the controller should not only review the probability of collection but also the number of bad debt requests submitted by the receivables clerk—a clerk who is pocketing customer checks may have submitted a disproportionate number of requests to convert accounts receivable into bad debts.

Closely Review Requests for Refunds

Requests for refunds may not be coming from customers; they may be coming from clerks within the company. To guard against this, a request for refund should be on the letterhead of the customer, which provides some guarantee regarding the source of the request. Of course, an enterprising employee could have letterhead created especially for the refund request, so a follow-up call to the customer is usually the best control over false refund requests.

Follow Up on Expected Miscellaneous Cash Receipts

Cash can be received for such miscellaneous items as the sale of company assets, insurance refunds, and legal settlements. The controller should track the dates and amounts of expected cash receipts and follow up on the cash settlements.

Otherwise, these receipts may be pocketed. There is an additional danger with miscellaneous cash receipts, because they tend not to arrive in the mail room and be logged in with the checks mailed from regular customers. Instead, legal settlements may be delivered directly to the in-house counsel, asset sale payments may go to the office manager, and insurance refunds may be mailed to the in-house risk manager. Because of the method of delivery, they may not be listed on the mail room control sheet and will thereby inadvertently skirt the company's cash control system.

Mutilate Voided Checks

The typical company voids a considerable number of checks each year. If not mutilated, these checks may be improperly cashed. A VOID perforation or a similar mark in writing will eliminate this risk.

Require Extra Authorizations for Bank Transfers

One wire transfer can easily move more money out of a company than thousands of separate petty cash thefts. Consequently, depending on the size of the transfer, multiple approvals may be required by senior management before any transfer is allowed to take place.

Mark or Mutilate Paid Invoices

It is not only likely but quite common for invoices to be paid more than once. To avoid this problem, invoices should be marked or mutilated in some manner that clearly identifies them as having been paid.

Lock Up the Signature Plate

If a signature plate is used to automatically affix a signature to checks, then it can be used to create improperly signed checks. Therefore, the signature plate should be locked up when not in use.

Fill in Empty Spaces on Checks

There is a danger that checks can be altered after they are signed so that the amounts paid are more than they should be. This can be avoided by having all checks printed by a computer, which automatically fills in all empty spots on the payment line with a meaningless symbol like an asterisk. Another control is to have the check signer seal the envelopes (an unlikely task!) and deliver them directly to the mail room.

Review Vouchers for Duplicate Submissions

Employees can resubmit vouchers for reimbursement in successive expense reports. This can be controlled by reviewing voucher dates for old vouchers that could have been submitted with previous expense reports and then checking previous reports for attached copies of those vouchers.

Update Signature Cards

All too often, people leave a company but can still sign checks months later because signature cards at the bank were not revised to reflect the departure. This can be controlled by immediately contacting the bank when an authorized check signer leaves the company to remove the person's signature from the list of valid check signers.

Do Not Disburse Large Sums with Cash

Large disbursements should be made with checks, since check transactions can be tracked with check registers and bank statements. Also, checks are safer than cash, since they are invalid until signed (whereas cash is already signed by the Secretary of the Treasury).

Do Not Allow Checks Made Payable to the Company to Be Deposited to Petty Cash

If checks made payable to the company are deposited into the petty cash account, the possibility exists that large quantities of cash can go into the petty cash account—and be withdrawn as well. The proper control is to allow only deposits of checks made payable to the petty cash fund.

Fill Out Petty Cash Vouchers in Ink

An employee can alter the amount on a petty cash voucher and pocket the difference. Requiring that vouchers be filled out in ink makes the forging task more difficult. The controller can then periodically review the vouchers for alterations.

Audit Petty Cash

Employees (primarily those in charge of petty cash) can take money from petty cash with the intention of returning it at a later date. This can be controlled by auditing petty cash without notice, so that missing money cannot be returned to the cash fund by the employee before the audit has been completed.

In summary, most cash-related controls cannot be changed without seriously weakening a company's control structure. However, shifting the company's cash activities into different modes of receipt or disbursement and types of cash can reduce the number of control steps used in a typical cash-related transaction.

QUALITY ISSUES

Accounting transactions are slowed considerably by the introduction of mistakes into the process, since the accounting staff must then track down the mistake, correct it, discern the reason for the error, and fix that problem as well. Thus, high-quality cash processing is defined as processing with the smallest number of mistakes. This section notes the most common cash processing mistakes and suggests ways to keep them from happening.

Incorrect Application of Cash Receipt against Receivable

A common error is to incorrectly apply received cash against a customer's accounts receivable record, requiring later research to correct the problem.

Incorrect Deposit Amount Recorded on Books

The cashier sometimes miscounts the amount of cash deposited and records an incorrect deposit amount. This not only requires later correction but also gives the company an inaccurate cash balance on its books, which may interfere with its cash management.

Premature Payment

An accounting clerk may enter an invoice date incorrectly, resulting in a one-time payment to the supplier on the wrong date. Alternatively, the clerk may incorrectly enter a supplier's payment terms into the supplier's master record, resulting in permanent payments on the wrong date.

Incorrect Amount of Payment

The incorrect amount may be entered for payment, resulting in later disputes with the supplier that causes strained supplier relations as well as extra work to solve the problem.

Duplicate Payment

A supplier invoice may be paid more than once. This causes additional work, since the supplier must be contacted and arrangements made for crediting the additional amount paid.

No Record of a Petty Cash Disbursement

Petty cash transactions may not have any supporting documentation, resulting in continuing petty cash shortages. If documentation is not included, the controller can only guess at the nature of the expense to be booked, or have a clerk contact all petty cash users to try to find some evidence of the expense for which the cash was used.

A number of techniques can be used to improve the quality of both incoming and outgoing cash transactions. To improve the quality of transactions, the controller does not want to add effort to the process by requiring supervisory reviews or data entries by a second person, since these methods are simply reviewing transactions for errors that have already been made. The best method is to keep the errors from occurring in the first place by reducing, eliminating, or simplifying the manual (error-prone) labor required to complete the processing.

Automate the Application of Cash to Receivables

An electronic funds transfer can include information about the allocation of the remittance to specific invoice line items. The processing of each EFT transaction should be fully automated, so that this incoming line item information can be offset against receivables in the recipient's computer system and automatically reduce the receivables balance. This is a good quality improvement measure because the errors caused by rekeying remittance information into the computer are avoided. A misapplied cash payment can cause considerable trouble to fix, since personnel from both companies must confer and review their records to determine where the problem occurred.

Have the Bank Record Deposits

Using a lockbox shifts to the bank the problem of incorrectly recording deposit amounts. Of course, the bank can also make mistakes, but a properly completed bank reconciliation each month (which should be completed anyway) should discover these errors.

Let the Computer or Folder Calculate the Payment Date

Nearly all computerized accounting systems will tell the user when to pay checks. For a manual system, incoming checks can be left in folders that are designated for specific weekly check runs—when the day's date equals the date on the folder, the check is paid.

Let the Computer Match Purchase Order Amounts to Invoice Amounts

A computerized accounting system can match supplier invoice amounts to the amounts listed on the initial purchase order and warn the accounting staff if the invoice amount varies from the originally agreed-upon price.

Only Pay the Amount on the Purchase Order

A new approach to making payments is not to bother with an invoice at all—just pay a supplier based on the unit price established in a purchase order (which should already be entered in the computer system) and base the quantity-delivered information on the amount recorded by the receiving department. This method has the advantage of using a price on the purchase order that may have been used multiple times in the past for previous deliveries, so that any errors in the purchase order price would probably have already been spotted and corrected. In contrast, an invoice is entered for every receipt from a customer, so the number of (error-prone) data entries are larger for invoices than for purchase orders.

Use EDI Invoices

Connect to key trading partners with EDI, so that information flows electronically from the computer systems of one company directly into those of the other company. Since the manual rekeying of data is eliminated, the quality of data received goes up. Of course, information may still be incorrect when it is first entered into the computers of one of the trading partners, but once the error is found and corrected, it never has to be manually re-entered, so the error rate goes down over time.

Let the Computer Spot Duplicate Invoices

When a payables department pays an invoice for a second time, it is probably not because the invoice became detached from the check and was paid again; rather, it is because the supplier sent an extra invoice (typically because the first invoice was not paid on time). To spot duplicate invoices, the company's accounting software should have a field in which the clerk must enter the supplier's invoice number; any duplicate invoice numbers will then be brought to the attention of management by the software. A further (and less common) control is for the software to match identical supplier invoice amounts and flag them for operator review.

Store Petty Cash in a Cash Register

Petty cash is frequently stored in a small lockbox or even in a desk drawer, and money can be taken from it with little regard for recording the transaction. By moving the cash to a cash register, it is locked up at all times, and a transaction amount must be printed on the register tape in order to open the cash drawer, which therefore leaves a trace of the amount of money taken, though not of its use.

Train with Audit Backup

Even the most experienced staff needs follow-up training to update them on the latest procedures needed to process transactions without making any errors. Since they usually don't believe they need the training (especially the more experienced staff), it helps to show them the results of internal audits that list the errors they have made while completing transactions.

In summary, the quality of cash transactions can be improved by using EDI transactions to avoid manual rekeying of information. In addition, the computer can automatically perform a number of quality checks, such as flagging invoices for payment only when due, noting differences between invoice amounts and purchase order amounts, and spotting duplicate invoice numbers and amounts. Also, errors can be avoided by simply eliminating the invoice (and the associated data entry) entirely and paying from the purchase order instead. Finally, ongoing training on how to complete error-free cash transactions should be conducted, along with feedback from the internal audit team regarding the number of errors occurring.

COST/BENEFIT ANALYSIS

This section contains an overview of how to prepare cost/benefit analyses for implementing lockboxes, area concentration banking, electronic data interchange, reduced payment approvals, and corporate credit cards. In these examples, expected revenues and costs are as realistic as possible.

Use a Lockbox

The banker of Mr. Longfellow, president of Poetic Moments (a publisher of poetry books), has convinced him to use a lockbox for processing incoming checks. As a long-time controller, you are skeptical that the lockbox is worth the effort. Your research reveals that the bank charges $30 per month to keep a lockbox open as well as a processing fee of 20 cents per check received. In a typical day, 80 checks would be received at the lockbox. An average daily deposit is $38,500, and the interest earned by Poetic Moments on its investments is 5%. The company would save one day of mail float by having checks sent to the lockbox. Internally, the mail room and accounting department staff are all hourly employees and earn an average of $13 per hour. It is company policy to send employees home when there is no work to do. The mail room staff person could be sent home a half-hour early if the incoming check control sheet no longer had to be prepared. Also, the cashier could save a half-hour per day by not having to prepare the deposit slip, and an accountant could save 15 minutes per day by not having to compare the control sheet to the deposit slip. Is it worthwhile to implement a lockbox?

Solution.   The annual cost is $360 for the lockbox fee and $3,520 to process a year's worth of receipts [20 cents × 80 checks/day × 220 business days], resulting in a total annual cost of $3,880. The total annual savings are as follows:

Additional Interest Earned

Average daily deposit	$38,500
Interest/day	× .014% [5% ÷ 365 days]
Total interest/day	$ 5.39
No. of business days/yr	× 220
Total interest/yr	$ 1,186

Labor Savings

Mail room	.50 hr
Cashier	.50 hr
Accountant	.25 hr
Total savings/day	1.25 hrs
No. of business days/yr	× 220
Total savings/yr	275 hrs
Hourly rate	× 13
Total savings/yr	$ 3,575

Thus, total annual cost is $3,880, and total annual savings are $4,761, resulting in savings of $881 per year. Based on the cost/benefit analysis, the lockbox should be installed.

Area Concentration Banking

The chief financial officer of Jiggers & Spinners, Inc., a manufacturer of fishing tackle, is concerned that the cost of compensating balances at subsidiaries is excessive. Currently, there are five subsidiaries around the country, each with $100,000 lines of credit, which on average require compensating balances of $20,000 in non-interest-bearing accounts. The CFO would like to know what change in cost would occur if cash were handled centrally through area concentration banking, thereby eliminating each subsidiary's compensating balance. You investigate the matter and find that money can be borrowed centrally with a 15% compensating balance, that the current investment rate is 6% for excess funds, and that the cost of a wire transfer between the accounts of subsidiaries and headquarters is $1 per transaction. You estimate there will be one wire transaction per business day per subsidiary. There are 220 business days in a year. What is the cost of decentralized financing compared with the cost of centralized financing, and what should you recommend to the CFO?

Solution.   By centralizing funds, the compensating balance on a total of $500,000 of lines of credit drops from 20% to 15%. This allows the difference, 5%, to be invested at the market rate of 6%, achieving savings of $1,500 per year. Offsetting the savings will be 220 wire transfers for each of the five companies, or 1,100 transfers. Since each transfer costs $1, the total offsetting cost is $1,100, which is less than the savings. The net gain is $400. Based on the cost/benefit analysis, the area concentration system should be implemented.

Electronic Data Interchange

Mr. Smedley, the founder of Deck Chairs, Inc., wants to tie his computerized order entry system to the payment system of the company's largest customer, K-Wal Department Stores. Currently, 50% of the company's business is with K-Wal. The company's revenues last year were $12 million. The average payment is $1,000 and takes 15 minutes to enter into the company's cash application system. The fully burdened cost of a company cash receipts clerk is $12 per hour. The company's internal auditor estimates that, based on selected samples, the company lost $20,000 last year because cash receipts were entered incorrectly into the system. The cost of purchasing EDI software is $3,000. You decide to use a value-added network (VAN) to transfer EDI transactions from K-Wal. The charge per transaction by the VAN is 50 cents. The MIS director estimates that programming the interface between the EDI software and the company's accounting software will take four months of a programmer's time. The fully burdened cost of the programmer is $38,000 per year. Should EDI be implemented?

Solution. The cost savings are from eliminating losses based on cash receipts being incorrectly entered (a loss of $20,000) and from eliminating the labor associated with the cash application.

Order Entry Labor Savings

Amount of K-Wal business/yr	$6,000,000
Average order size	÷ $1,000
No. of K-Wal payments/yr	6,000
Time to enter one receipt	× .25 hr
Time/yr to enter K-Wal receipts	1,500 hrs
Cost of receipts clerk/hour	× $12
Total savings/yr	$ 18,000

Cost of EDI Implementation

Cost/yr of programmer	$38,000
Four months' work	÷ 3
Total programming cost	$12,667
Cost of EDI software	3,000
Total EDI implementation cost	$15,667

Cost of EDI Transactions

No. of K-Wal orders/yr	6,000
Cost/EDI transaction	× $.50
Cost/yr of transactions	$3,000

Since the total cost of $18,667 and the total benefit is $38,000, the cost/benefit analysis indicates that an EDI system should be implemented.

Reduce the Number of Payment Approvals

A key supplier has called the president of Tender Tissues, a maker of non-allergenic tissue paper, and complained that payments are extremely late. The president informs Mr. Underling, the controller, that this is unacceptable. Investigation by Mr. Underling reveals that several suppliers have been paid so late that they have forced Tender Tissues to accept C.O.D. terms, which carry a 5% surcharge; these payments account for 10% of the company's total annual purchases of $2,600,000. Internally, payments are made as soon as approved invoices are returned by supervisors to the accounts payable department. However, the time required to obtain signatures on invoices can take up to a month, since supervisors travel constantly. The solution appears to be the elimination of supervisor approval of supplier invoices and the strict enforcement of purchase requisition sign-offs. To enforce purchase requisition sign-offs, Mr. Underling estimates that a staff person (annual salary $32,000) must be assigned to the task full-time for four months. In addition, an internal auditor (annual salary $36,000) must review the purchase requisition controls for two weeks every year to report on any deviations from the control system. Is eliminating control over invoices worthwhile to the company?

Solution. The costs and benefits of reducing the number of payment approvals are as follows:

Cost of Enforcing Requisitions

Cost/yr of staff	$32,000
Four months' work	÷ 3
Cost of staff	$10,667
Cost/yr of auditor	$36,000
Two weeks' work	× .038
Cost of auditor	$ 1,368
Total salary cost	$12,035

Benefit of Stopping C.O.D. Payments

Purchases/yr	$2,600,000
C.O.D. payments	× 10%
C.O.D. payments/yr	$ 260,000
C.O.D. surcharge	× 5%
C.O.D. surcharge/yr	$ 13,000

Since the total cost of eliminating invoice approvals is $12,035 and the benefit is $13,000, the cost/benefit analysis indicates that invoice approvals should be eliminated.

Use Corporate Credit Cards

As the controller of Wimpy Spinach factory, you are concerned that too much accounting effort is being expended on paying for small-dollar items. Research reveals that 80% of all checks are for less than $150 dollars and of these one half are made out to three companies, dealing in office supplies, shop supplies, and janitorial supplies. None of these suppliers allow purchases with credit cards, but the company's business can be switched to three other suppliers who *do* accept payment with credit cards but whose products are on average 1% more expensive than those of the current suppliers. Total company expenditures per year are $1,500,000, and on average 1,800 checks are written per year. A recent activity-based costing study has noted that the cost to process a check is $8.50. Is it worthwhile to purchase small-dollar items with credit cards?

Solution.   The costs and savings in this example are as follows:

Cost of Multiple Check Payments

Total expenditures/yr	$1,500,000
Small-dollar payments	× 80%
Small-dollar payments/yr	$1,200,000
Payments to three suppliers	× 50%
Payments/yr	$ 600,000
Cost to change suppliers	× 1%
Total change cost/yr	$ 6,000

Benefit of Using Credit Cards

Checks written/yr	1,800
Small-dollar checks	× 80%
Small-dollar checks/yr	1,440
Checks to three suppliers	× 50%
Checks/yr	720
Cost per check written	× $8.50
Total savings from not writing checks	$6,120

With costs of $6,000 and benefits of $6,120, the cost/benefit analysis indicates that it is profitable to purchase small-dollar items with a corporate credit card, thereby reducing the number of payments to 12 credit card checks per year.

In summary, the preceding cost/benefit examples can be used to develop real-life analyses, especially in terms of the line items used. The dollar amounts of the costs and benefits in these examples were taken from the author's experience; however, readers' cost and benefit projections should use amounts derived from readers' specific situations, not from these examples.

REPORTS

A company usually already has cash reports related to cash receipts, cash expenditures, and actual bank balances versus required balances. The changes noted in this chapter do not require changing any of those reports. However, some additional information is needed, depending on the changes being implemented.

Electronic Funds Transfer

When cash is transferred by electronic funds transfer (EFT), the specific receivables line items that add up to the total of the cash transfer must be listed. This information is used by the company to offset the receivables balances in its records. It usually arrives in the form of an EDI transaction.

Corporate Credit Cards

Credit card companies should submit monthly reports that detail not only what items were purchased but also who purchased it. This report allows the company to allocate costs to the correct general ledger accounts and to list purchasing activity by individual user, so that potential abuses can be spotted after the fact.

Cash Volume

The company should create a report that sorts cash receipts and expenditures by volume of transaction (see Tables 1.3 and 1.4). This is used to highlight the company's top trading partners in terms of cash transactions and tells the controller which trading partners might be prospects for links with EDI. The existing cash forecast will become more accurate when EDI is used, because the company knows precisely when funds will be transferred to suppliers or from customers, thereby eliminating the variability of the mail float and check clearing float.

In short, existing cash reports are not altered by the revised cash system, but new reports can be created that give details about purchases using corporate

TABLE 1.3   Cash Volume by Customer, Sorted by Sales Volume

Customer Code	Customer	YTD Cash Received	YTD Sales
PARROT	Parrot Ink Supply Company	$3,000,009	$3,215,000
COLORS	Colors Are Us	2,100,452	2,467,000
MIXMAT	Mix 'n Match, Inc.	1,111,109	1,111,109
B&W	Black and White Partners	902,113	942,333
MAUVE	Mauve Associates	833,001	844,909
SKYBLU	Sky Blue Designs Company	641,056	700,319
DESERT	Desert Sand Associates	421,098	601,001
REDBRI	Red Brick Construction Company	308,722	549,765

TABLE 1.4   Cash Volume by Supplier, Sorted by Order Volume

Supplier Code	Supplier	YTD Cash Paid	YTD Ordered
DUMONT	DuMont Chemicals	$4,003,622	$4,101,022
ACCESS	Access Chemicals	2,098,555	2,222,098
INTERF	Interferon Petroleum	1,993,005	2,100,001
INTKER	Interstate Kerosene Supplies	1,702,113	1,803,000
INTLOX	International Lox Distributors	1,508,422	1,599,044
MIXING	Mixing Equipment Company	1,109,888	1,208,900
HRDUNN	H.R. Dunn Manufacturing Supplies	992,073	1,000,030
FOSS	Foss' Chemical Nullifiers	840,156	913,498

credit cards or about high-volume trading partners who should be converted to EDI transactions.

TECHNOLOGY ISSUES

Several recent innovations can improve the cash-processing function.

Corporate Purchasing Cards

This is a new use of familiar technology—the credit card. In essence, a company contracts with a credit card company to provide credit cards to key employees who are responsible for purchasing large quantities of low-priced goods. The purchasing limits on the cards are set low to prevent overspending abuses. The cards can also be limited to purchases from specific suppliers, thereby ensuring that only certain classes of goods are purchased. Some credit card companies send itemized billing statements back to the companies for easy processing against appropriate account numbers. Corporate purchasing cards are useful for reducing the time spent creating purchase requisitions and purchase orders for small-dollar items as well as the accounting time spent processing many small-dollar invoices.

Electronic Data Interchange

This can no longer be considered a new technology, for EDI has been in use since the 1970s. However, it has been used most frequently for processing invoices and purchase orders, not cash transfers. As it relates to cash, a customer sends an EDI transaction to a supplier, advising of a payment authorization going to the bank at the same time via electronic funds transfer. The EDI transaction includes an itemization of what the cash transfer is paying for, so that the supplier can use the EDI information to charge the cash receipt off against specific receivables line items. The transaction is helpful in reducing receivables balances earlier than with payment by mail, thereby reducing the number of collection calls made for overdue receivables. Also, if a large enough discount is used to entice the customer, a supplier can request same-day payment for delivered goods, thereby reducing working capital requirements. EDI can also be used to instruct a bank as to the exact date on which funds will be released to a supplier. This assists both the customer and the supplier in their cash forecasting, since they know the precise date on which money will move into or out of their accounts.

Bar-Coded Prices

One of the controls used to prevent theft of cash is to involve more than one employee in a cash transaction, so that collusion is required to steal the cash. For example, in a retail clothing store, the sales clerk gives a sales ticket to the customer, who takes it to a cashier to pay for the clothes. However, to improve the speed of the transaction, the controller may want to concentrate both functions in one person. A means of centralizing the functions while maintaining control is to bar-code the price and allowing the sales clerk to scan the bar code into the cash register. The preprinted bar code does not allow the clerk to enter any amount besides the amount on the tag, thereby avoiding the problem of having the employee enter a different amount into the cash register and pocketing the difference.

Positive Pay

A very recent innovation that slows down the cash-processing function but nearly eliminates check fraud is positive pay. In this process, the company sends the bank a daily list of checks issued. This allows the bank to reconcile the company's list of checks with the checks presented for payment and to spot any discrepancies. The bank can also reverse the process by making available a list of each day's cleared checks, which the company can then reconcile to its list of printed checks.

Chemically Treated Checks

Though it has no effect on the speed of cash transactions, a recent innovation is to purchase check stock that has been treated with special chemicals that deface the check if it is chemically tampered with or photocopied. This is an improved control over standard check stock, since the bank will realize that a check has been tampered with if someone tries to cash a disfigured check.

In summary, several technological advances in the last two decades contribute to quicker cash transactions and more control over cash transactions. These innovations are essential to the revised cash receipts and disbursements systems outlined in this chapter.

MEASURING THE SYSTEM

Different types of measurements are needed to determine the efficiency of the streamlining suggestions mentioned in this chapter for improving the speed of cash flow, reducing the transaction-processing work of the accounting staff, and reducing the time required to complete a transaction.

Improving the Speed of Cash Flow

This item relates to the use of lockboxes, area concentration banking, and zero cash balances. At a high level, the easiest measurement of faster cash flow is days of receivables outstanding. However, this measurement includes other factors besides faster cash flow, such as the creditworthiness of customers. Instead, to measure the speed of cash flow, the controller can use the customer's check date as a baseline and track how quickly this money clears the bank from the time it was sent to the company by the customer. This tracking can be very time-consuming, since the check date and clearing date are not listed in the databases of most computerized accounting systems. The controller can also track the percentage of cash in interest-bearing accounts, which shows how well (or not) money is being kept out of non-interest-bearing accounts and concentrated into interest-bearing investments.

Reducing the Transaction-Processing Work Load

This item relates to the use of petty cash, corporate credit cards, and EDI transactions. The controller should measure the time required by each employee to perform a task, both before and after one of the recommended changes has been implemented. This is not an ongoing performance measurement but rather one that may be periodically measured during an activity-based costing review or by internal audit staff during a scheduled review.

Reducing the Time Required to Complete a Transaction

This item relates primarily to the elimination of invoice approvals. The measurement needed is to track the number of days between the date an invoice was received in the mail by the accounts payable department and the date all paperwork related to the payable (including the invoice, purchase order, and receiving documentation) was completed. By eliminating the approval of invoices, this measure should drop by at least one day and possibly by as much as several weeks (especially in the case of a multioffice company with a centralized accounts payable department).

In summary, most of the performance measurements advocated in this chapter are not the standard ones included in typical accounting systems; the controller must decide if it is worthwhile to assign staff to track this information. A compromise is to track this information on a sample basis periodically to determine if performance is changing appreciably over time. This may be a task for the internal audit department.

IMPLEMENTATION CONSIDERATIONS

A number of problems relating to change management may arise when revised cash receipts and disbursements systems are installed (see Chapter 11). The following problems can be expected when implementing revised cash systems.

Lockbox System

When customers are asked to start sending their checks and remittance advices to a lockbox, at least half of them on average will not switch on the first request. Instead, the controller must be prepared to wage a year-long campaign of letter writing, phone calls, and contacts by company management with the management of the customer before the company's address will be changed in the databases of all customers. Reasons for this are misroutings of the mailed request at the customer location, intransigent staff, or a supplier management that realizes a lockbox system will reduce its mail float. Also, the controller must be sure to change the mail-to address on the company's invoices or affix a new address label onto the old invoice form.

Area Concentration Banking

When a multilocation company switches to area concentration banking, the biggest problem is getting the banks that service each location to regularly shift funds into a central account. This is partly due to the unwillingness of a bank to shift money (on which it is earning interest) immediately into the account of another bank as well as to the lack of daily account tracking and wire transfer capabilities at some of the smaller banks (though this is increasingly rare). The solution is to use one of the national banks that have branches or affiliates in most parts of the country. These banks actually encourage area concentration banking, since they garner fees no matter where the funds go within their far-flung systems.

Electronic Data Interchange

It is relatively easy to install EDI as a stand-alone system. However, if the company wants to have EDI cash transactions automatically offset accounts receivable records in the company's accounting database, an interface between the received transaction set and the accounting system must be constructed. Such an interface can be quite difficult to construct. There is no way to get around this problem, but the controller should consult the management information systems (MIS) manager to get an estimate of the time and cost required to complete the work. This information should be obtained early in the project, so that the company will not have erroneous expectations for the completion date.

Reducing the Number of Payables Approvals

A memo from management can easily stop the additional approvals that are frequently used in the cash disbursements process. However, the reason so many approvals were originally required was that not all expenditures were preapproved during the purchase requisition phase. Many companies have a leaky purchasing system, in which items can be ordered by nearly any employee without the control imposed by a purchase requisition. Consequently, before eliminating management approval of supplier invoices, the controller should work with the internal audit department to stop the leaks in the purchasing system. This means tracking down who ordered items without a purchase requisition and reprimanding them, informing all company suppliers that only parts and services accompanied by purchase orders will be paid for, and continually reviewing purchase requisitions to ensure that the appropriate level of authority has signed off on every requisition or purchase order.

Petty Cash

When the controller allows large numbers of small disbursements to be paid for out of petty cash, it is likely that the number of disbursements will be unaccounted for, since the petty cash clerk is bound to make mistakes as the volume of transactions grows. Two ways to avoid this problem are to install a cash register for petty cash (which provides a record of disbursements with the cash register tape), and to provide extra training to the petty cash clerk to reinforce the petty cash accounting procedure. Periodic reporting of petty cash errors by the internal audit department will reinforce the need for care when processing petty cash transactions.

Computerized Accounting System

All the problems associated with any major system changover can be expected when converting from a manual to a computerized accounting system. Of particular concern to the controller is the integrity of the accounting data. With that goal in mind, the controller should pay particular attention to the conversion of the manual database to the computer database, including testing the converted data for accuracy. To keep the data accurate, the controller should enforce a definite conversion date, so that employees are not using old data from the manual system when they should be using and modifying the data in the new database. Finally, proper training in the use of the new system will help to keep the data from being corrupted by improper use.

In short, implementation problems may occur when switching to the faster cash accounting transactions outlined in this chapter. These problems are caused by intransigent employees and suppliers, conversions to new computer software, and tightening the new controls that need to replace the old ones.

SUMMARY

The maze of controls that surround the processing of cash are there for a good reason—to prevent the loss of cash by ineptitude or embezzlement. Except for recommendations to eliminate the review of supplier invoices and checks to suppliers, this chapter has not advocated getting rid of any of the traditional controls. However, cash can be converted to other forms that require fewer controls, or it can be received through alternative channels that require fewer controls. Consequently, this chapter has focused on changing the type of cash to electronic (electronic funds transfer, credit cards) and on having cash enter and depart the company's bank without flow through the company. This chapter has also suggested disbursing high-volume, small-dollar payments through the petty cash fund in order to avoid the many controls surrounding check disbursements; the same goal can be achieved by reducing the number of cash transactions into a small number of large payments with corporate credit cards. By using these various means to reduce the number of controls surrounding cash receipts and disbursements transactions, the speed of accounting for cash transactions will increase.

REFERENCES

Arnette, Denise A. "Banks Turn Imaging into Speedy New Delivery Products." *Corporate Cashflow,* January 1994, pp. 12–13.

Forman, Larry, and Theresa Paquin. "Tapping Technology to Make Money." *Corporate Cashflow,* June 1992, pp. 23–28.

Holland, Kelly. "Bank Fraud, the Old-Fashioned Way." *Business Week,* September 4, 1995, p. 96.

Queree, Ann. "It's a Buyer's Market." *Corporate Finance,* March 1994, pp. 14–18.

Saunders, Gary J., and Ruth E. Saunders. "Cash Forecast Template." *The Financial Manager,* Sept./Oct. 1989, pp. 40–45.

Whittington, O. Ray, and Kurt Pany. *Principles of Auditing.* Chicago: Irwin, 1995.

Willson, James D., James P. Colford, and Janice M. Roehl-Anderson. *Controllership: 1995 Cumulative Supplement.* New York: Wiley, 1995.

Willson, James D., Janice M. Roehl-Anderson, and Steven Bragg. *Controllership.* 5th ed. New York: Wiley, 1995.

2

SALES AND ACCOUNTS RECEIVABLE

This chapter discusses the entire sales cycle, since it is difficult to separate accounts receivable from the whole sales cycle. A typical sales and receivables system has an extraordinary number of controls related to separating various duties in the process, all designed to make sure that no one can manipulate reported sales volumes or steal incoming payments for accounts receivable. As transactions wend their way from one person to the next, it is very easy for the transaction to be halted for any number of reasons, each of which contributes to slowing the transaction. The only way to increase transaction speed appreciably is to consolidate the control functions with fewer people, but this interferes with the desired separation of duties. So how does one reduce the number of controls safely in order to increase the speed of the accounting transactions? This chapter describes a typical set of sales and accounting transactions and shows how to improve the speed while maintaining proper control by using entirely new transaction methods.

The modified system is then reviewed for control weaknesses and ways to improve the quality of its output. A detailed cost/benefit analysis is presented, which can be used as a model to determine if a company's systems should be converted to the modified system. The chapter concludes with a discussion of reports to be expected from the modified system, potential implementation problems, the effect of new technology on the system, and ways to measure its performance.

CURRENT SYSTEM

This section describes a typical sales and receivables transaction, and notes the flow of information between employees. A value-added analysis chart shows how much time is taken by each of the steps in the transaction. Later sections recommend various techniques for speeding up the process.

The sales transaction begins when a customer purchase order arrives in the company mail room. Unlike cash, purchase orders are not logged in by the mail room staff. Instead, they are accumulated and placed in the mail slot of the order-processing clerk, who is normally located in the sales department.

The order-processing clerk uses the information on the purchase order to create a sales order. The sales order is simply a revised form of the purchase order, restated so that, for example, the company-assigned part numbers for the parts being ordered are listed. The sales order also lists check-off spaces for such items as credit approval, entry into the production schedule, and shipping.

The sales order then moves to a credit analyst, who is usually located in either the treasury or the accounting department. If a customer is already set up in the system with an unused credit line, the sales order is approved and quickly moved to the next step. If the credit line has been used or if the customer is new, the credit analyst must review the customer's financial condition by checking credit reports and financial statements. This step may require some delay while the customer sends information about itself to the company. If the credit information is good, the analyst assigns a credit amount to the customer, approves the sales order, and passes it along to the production control department.

The order enters into the production and warehousing area, which falls outside the analysis of this chapter. After the parts have been picked from stock, ordered and received from a supplier, or produced, they are sent to the shipping department.

The shipping department reviews the sales order to ensure that the credit analyst has approved the order. If so, the parts are shipped. If not, the parts stay in the shipping department until credit approval has been given. Realistically, credit approval should not have to be checked in the shipping department, because the sales order should not have been released to the production department unless credit was approved. However, many companies perform this last-minute check to ensure that goods are not sent to a customer that may not be able to pay for them.

When the order is ready to ship, the shipping department creates a bill of lading to go with the shipment. The shipment is listed in a shipping register that shows the sales order number, freight carrier name, ship date, and customer name. A copy of each day's shipping register is then forwarded to the accounting department, which compares the shipping register to the sales order, customer purchase order, and any change orders. The accounting department applies any discounts based on purchasing volumes or special deals, and issues the invoice. If shipments are related to a long-term contract or are being sent to a government entity, a review of the related contract for pricing terms may also be necessary. It is best not to have the sales department issue invoices, since the sales department may be tempted to improperly issue invoices to increase reported sales levels.

After the invoices are created, a daily batch total is entered into a sales journal, which is later summarized and entered into the general ledger for the accounting period. A typical sales transaction is shown in Figure 2.1.

The sale now becomes a receivable. Many companies include it in a monthly statement that is sent to all customers. After the due date has passed, the company's collections clerk calls the customer's payables department to ascertain the status of payment for the invoice. The collections clerk may have to send additional information to the customer, such as proof of shipment or a copy of the invoice. Dunning letters may also be sent to the customer.

After the customer pays for the invoice, the received cash is used to reduce the balance of the customer's account. Normally, the customer payment includes a remittance advice listing all the invoices being paid. This information is used to reduce specific line items in the customer's account record. If the payment detail does not accompany the payment, the customer's accounts payable department must be called to find out what is being paid for by this payment. A typical accounts receivable process is shown in Figure 2.2.

A mismatch that commonly occurs between the accounts receivable department and the credit collections staff is that the accounts receivable records may not be updated quickly enough, resulting in collection calls to customers who have already paid the company.

A large amount of paperwork is involved with sales and receivables transactions. The customer purchase order is rekeyed into a sales order, to which are added a bill of lading and an invoice. A credit memo may be added to the transaction as well as a notice of back-order. The customer may later be sent a monthly statement of outstanding invoices as well as various dunning notices. This flow of paper is noted in Figure 2.3.

A review of the flowcharts reveals a great deal of movement of paperwork between various departments. The move and wait times thus introduced into the process greatly slow it down. Also, every time papers are moved, the potential exists for loss or misinterpretation.

The value-added analysis shown in Table 2.1 lists each step in the process and the time required to complete each step. A value-added item is considered to be one that brings the sales/receivables transaction closer to conclusion. This analysis shows optimistic wait times while paperwork waits in employees' mailboxes. A statistical analysis of an actual situation will reveal a small number of much longer wait times (because of employee absence) that will appreciably affect the completion time of the transaction.

Table 2.2 provides a summary of the value-added analysis. It shows that only one-quarter of the steps bring the sales/receivables transaction closer to conclusion; the remaining activities are related to moving paperwork from person to person or making copies for the files. In terms of time required, the value-added

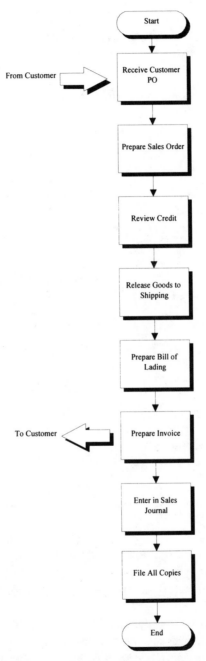

FIGURE 2.1 Typical sales transaction.

FIGURE 2.2   Typical receivables process.

steps can be concluded in 1.1 hours, while the moving, waiting, and non-value-added portions of the transaction take up nearly 2.5 business days. In short, the actions needed to conclude the transaction make up only a small part of the total process.

When alternatives for speeding up the process are discussed in the next section, the proposed changes are not confined to the move, wait, and non-value-added steps. Even the value-added steps are reviewed for possible changes or elimination.

In summary, the traditional sales and accounts receivable process is burdened by a large number of controls that require paperwork to move through many

FIGURE 2.3   Paperwork in the sales/receivables process.

TABLE 2.1   Sales/Receivables Value-Added Analysis

Step	Activity	Time Required (Minutes)	Type of Activity
1	Receive customer purchase order (PO) in mail room.	1	Value-added
2	Wait—accumulate in pile until mail finished.	15	Wait
3	Carry to mail slot.	1	Move
4	Place POs in employee mail slots.	1	Non-value-added
5	Wait for employee to check mail slot.	120	Wait
6	Carry PO to desk.	1	Move
7	Prepare sales order (SO).	10	Value-added
8	Wait while other POs in batch are converted to SOs.	60	Wait
9	Move to copier machine.	1	Move
10	Make copies of PO.	1	Non-value-added
11	Move back to desk.	1	Move
12	File copy of PO.	2	Non-value-added
13	Carry PO and SO to mail slot.	1	Move
14	Place PO and SO in employee mail slot.	1	Non-value-added
15	Wait for employee to check mail slot.	120	Wait
16	Carry PO and SO to desk.	1	Move
17	Review customer credit records.	10	Value-added
18	Call customer to request information.	5	Value-added
19	Wait for customer information to arrive.	480	Wait
20	Review customer-supplied information.	10	Value-added
21	If acceptable, sign off on SO.	1	Value-added
22	Move to copier.	1	Move
23	Make copies of PO and SO.	1	Non-value-added
24	Move back to desk.	1	Move
25	File copies of PO and SO.	2	Non-value-added
26	Move to mail slot.	1	Move
27	Place PO and SO in mail slot.	1	Non-value-added
28	Wait for employee to check mail slot.	120	Wait
29	Move PO and SO to desk.	1	Move
30	Write down order on production schedule.	5	Value-added
	[Production occurs and product is shipped.]		
31	Move to copier.	1	Move
32	Make copies of PO, SO, and bill of lading.	1	Non-value-added
33	Move to desk.	1	Move
34	File copies of PO, SO, and bill of lading.	1	Non-value-added
35	Move to mail slots.	1	Move
36	Place PO, SO, and bill of lading in mail slot.	1	Non-value-added
37	Wait for employee to check mail slot.	120	Wait
38	Move PO, SO, and bill of lading to desk.	1	Move
39	Prepare invoice.	5	Value-added

TABLE 2.1 *(Continued)*

Step	Activity	Time Required (Minutes)	Type of Activity
40	Wait for other invoices in batch to be prepared.	60	Wait
41	Move to copier.	1	Move
42	Make copies of PO, SO, bill of lading, and invoice.	1	Non-value-added
43	Move to desk.	1	Move
44	File copies of PO, SO, bill of lading, and invoice.	1	Non-value-added
45	Move to mail room.	1	Move
46	Mail invoice.	1	Value-added
[Payment from customer is received.]			
47	Mail room staff withdraws payments from letters.	1	Value-added
48	Wait for all envelopes to be opened.	15	Wait
49	Mail room staff records check information on register.	5	Value-added
50	Move checks to mail slot.	1	Move
51	Place checks in mail slot.	1	Non-value-added
52	Wait for employee to check mail slot.	1	Wait
53	Move to desk.	1	Move
54	Contact customer about incorrect or unidentified payment amounts.	10	Value-added
55	Post payment to customer account.	5	Value-added
56	File invoice.	1	Non-value-added

departments, introducing the risk of lost or misinterpreted information. The following section reviews ways to bypass several traditional accounting controls, thereby allowing faster transactions to occur. In addition, several methods are discussed for reducing the risk of losing or altering key information related to sales and accounts receivable.

TABLE 2.2 Summary of Sales Receivables Value-Added Analysis

Type of Activity	No. of Activities	Percentage Distribution	No. of Hours	Percentage Distribution
Value-added	13	23%	1.10	6%
Wait	10	18	18.52	91
Move	19	34	.32	2
Non-value-added	14	25	.27	1
Total	56	100%	20.21	100%

REVISED SYSTEM

To complete the sales and receivables cycle requires a considerable amount of time and the participation of a great many employees. This section reviews ways to speed up the cycle by cutting the paperwork move and wait time, eliminating some staff involved in the process, and eliminating steps from the process.

One solution to speeding up the sales and receivables process seems simple enough—just concentrate all the functions with one person to avoid the move and wait intervals, and have that person complete all the steps. Such a system is shown in Figure 2.4.

However, concentrating all the steps with one employee violates a number of key control issues. For example, a salesperson assigned responsibility for all of these tasks might grant credit to all customers in order to maximize commissions. Or, someone who prepares and mails the invoice should not log the payment into the accounts receivable system, since a dummy invoice for an excessive amount could be sent to the customer and the excess amount pocketed upon receipt. Finally, there is the risk that a shipment could be sent to a fake company owned by an employee who can authorize credit for the shipment. In short, control problems do not allow the controller to centralize all sales and receivables functions with one person, despite the obvious efficiencies that would result. Other means must be found to improve the transaction speed.

One method is to transfer sales and receivables information between employees *using electronic mail.* Though this is rarely done, it would be possible to complete a processing step and then move the entire set of electronic documents into the electronic mailbox of the next person who is supposed to review it. For example, as shown in Figure 2.5, a customer purchase order can be physically converted into a sales order by an order entry clerk, and all transaction steps from that point on can be transferred from employee to employee by electronic mail. This allows the staff to save documents electronically and eliminates the move time associated with carrying documents between employees. However, just because a person's electronic mail is accessible from that person's computer terminal does not mean that it will be accessed any more quickly (though the wait time is generally reduced) than would be the case if the materials were left in a company mail slot. Also, a problem with using electronic mail to transfer documents between employees is that sometimes an employee must attach an approval to the document; this requires an access password that must be changed regularly, or else it may be copied and used by an unauthorized person. In short, electronic mail introduces some automation to the sales/receivables process but does not change the basic process.

Another approach to speeding up the process is *preapproving customer credit.* Nearly all companies allow their salespeople to pursue customers and procure

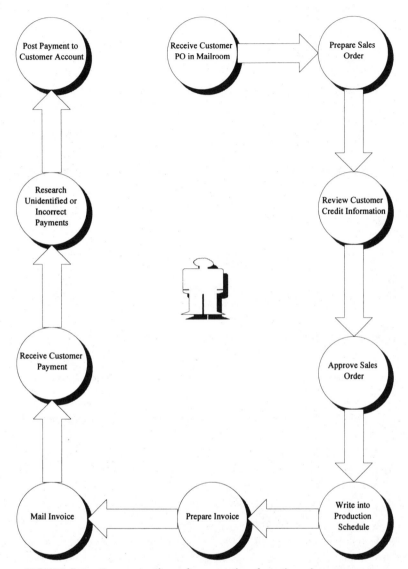

FIGURE 2.4   Concentration of accounting functions in one person.

orders before checking on the credit of the customers. This leads to excessive pressure on the credit analyst, who must now collect credit information about the customers and make decisions about the amount of credit (if any) to be extended. Meanwhile, pressure is brought to bear on the credit analyst by the salesperson, since the latter wants to hurry through the purchase orders that translate into commission income for the salesperson. Overall, the situation causes orders to be delayed, since it may take more than a day to procure credit information and

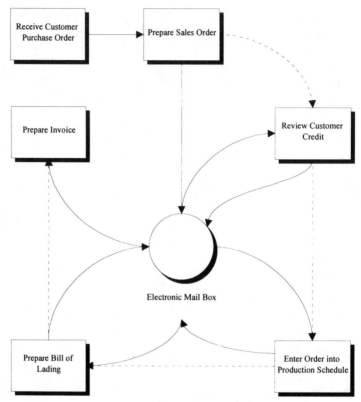

FIGURE 2.5   Use of electronic mail in the sales transaction.

reach a decision, and sometimes leads to bad credit decisions, since the sales department is determined to book the orders. An alternative approach is to team
with the sales department in advance, go over the credit records of potential customers, and assign them credit before the salesperson ever makes a sales call.
This speeds up the sales/receivables process in two ways: first, it takes the credit
information collection step out of the process (though incoming purchase orders
must still be checked against a list of preapproved customers). Second, the new
procedure keeps the company away from potential customers with bad credit
records, which reduces the time required to collect receivables from customers
that are unable to pay. Since credit status changes over time, it is important to review the records of current customers on a regular basis. In short, by preapproving the credit of target customers, the company can reduce the time required to
process a sales order and the effort needed to collect overdue receivables.

Another method for reducing processing time is *handing the customer an invoice
when the product is delivered.* This eliminates the time needed to create an invoice
at the company location, mail it to the customer, and wait for the mail to be delivered to the customer. The best approach is to have the invoice information stored

on a portable computer; after the customer's receiving department inspects the shipment, the delivery person can input any changes into the computer and print out an adjusted invoice on the spot. Alternatively, the delivery person can hand over a diskette that contains the invoice information. Since most customers are not equipped to handle electronic invoices, a paper-based invoice is the more common approach. This is only a viable option when the company has its own in-house freight carrier, since third-party carriers are not authorized to modify and deliver invoices.

A way to reduce the work load of the accounts receivable department is *eliminating the mailing of month-end statements*. Many accounts payable departments do not have the time to review these statements and simply throw them out without comparing the information on them to the records. Furthermore, a periodic statement is not an invoice, so a customer cannot pay the company from the statement anyway—the customer must call the company and request an invoice for any items listed on the statement that are not in its records. Very few accounts payable departments have the time to follow up on statements in this manner. Thus, it is best to eliminate the printing and mailing of periodic customer statements.

The dunning process that is an integral part of the accounts receivable department can be speeded up by *immediately updating receivables records*. When cash is received, a primary task is to log the cash receipt against the receivables record. This information must be readily accessible to the employees who issue dunning notices or make credit collection calls. Otherwise, they will waste time trying to collect money that has already been received; this slows down the overall collection process because a fixed amount of employee time is being spread over too many collection calls and letters. The best way to ensure that receivables records are immediately updated is emphasizing the recording of cash receipts as top-priority and using a computerized accounting system that updates receivables records in real time. This combination will reduce the work load of the company's collection personnel and speed up the collection of receivables.

Eliminating duplicate *receivables and collection systems* can also speed up transactions because payments sent to the wrong system require time-consuming reconciliations between systems. One study discovered that the average company has 4.6 accounts receivable systems and 3.6 credit and collections systems per billion dollars of revenue.[1] Clearly, consolidating these systems into a single system will improve efficiency.

Another way to improve the speed of collections work is *creating an automated collection tickler file*. This is a computer program that is tied to the due dates on the company's accounting software and tells the collections department when an invoice is overdue. The collections staff can type contact information

---

[1] Hackett Group, *AICPA/THG Benchmark Study: Results Update and Analysis,* 1994.

into the tickler record, which is stored and available the next time the record is used by someone. Each record in the tickler file should have a default number of days between customer contacts, so that the delinquent account will be brought to the attention of the collections clerk on a regular basis. The clerk should also have an option to type in the number of days before the next contact, in case a longer or shorter period than the default amount is more appropriate. This system allows the collections employees to make more calls, since they are not wasting time searching for information about overdue accounts—the correct information is presented to them in the tickler file. The tickler file speeds up the collection process by eliminating the work to relocate and refile, collections papers.

A way to speed up the invoicing process is *sending invoices to customers using electronic data interchange (EDI)* (see Chapter 8). Briefly, the company sends an electronic invoice to a central clearinghouse via modem, where the invoice is stored. The customer accesses the clearinghouse regularly by modem and can receive the electronic invoice at that time. Since an acknowledging EDI transmission is sent back to the company by the customer, the company will know that the invoice has not been lost. The mail float associated with a mailed invoice is also eliminated, since the transmission can reach the customer in moments (if there is a direct connection). The real improvement from using EDI has little to do with the method of transmission, however. Rather, the sales/receivables cycle is speeded up by linking it to the company's shipping system, so that invoices are prepared and sent automatically. This eliminates manual invoice preparation and thereby shrinks the sales/receivables transaction cycle. In short, EDI allows the company to eliminate a processing step and delete the mail time normally needed to transmit an invoice to the customer.

The customer's accounting department may have an invoice recorded in the accounting database but may not have the invoice. This is because invoices are usually sent out to supervisors for approval, and an invoice may be lost by a supervisor. If an invoice is lost, the customer's accounting department has no way of knowing about it until a collections call is made for an overdue invoice; by the time a replacement invoice has been sent, even more time will have gone by. To avoid this problem, the company's collections staff can *call customers' payables departments prior to the invoice due date* to ensure that the invoice is approved for payment. This allows the collections staff enough time to send out a replacement invoice and still collect the payment in a timely manner.

A company that ships to a customer on a just-in-time basis may be able to *eliminate all invoices* by having the customer pay the company based on the number of the company's products used in the customer's final product. In other words, the company ships a small number of its products to the customer often. The company is the only supplier of that product to the customer. When the product arrives at the customer's production facility, it is immediately included in the production process. When the customer's product is completed, the customer's accounts payable system can calculate the amount due to the company based on the number of its parts in the customer's finished product. This system has no need for invoices. Instead, payments will be sent to the company periodically; the remittance advice attached to the payment will list the customer's purchase order number as a reference instead of the usual invoice number—the company's accounting systems must be reprogrammed to handle this change of information. In short, the invoice creation process can be avoided entirely if the company's customers pay the company based on their production records.

In summary, the sales and receivables transaction cycle can be speeded up by cutting out functions, reducing the causes of slowdowns in the process, handling certain processes in advance, and using technology to speed up the flow of information. When all these methods are combined, transactions will flow through the company much more swiftly. A flowchart of the revised sales/receivables cycle is shown in Figure 2.6. The following section reviews the control issues associated with these changes.

CONTROL ISSUES

There are many control issues related to the sales and receivables transaction cycle. For example, if the controller were to eliminate controls related to granting credit to customers, the risk of shipping products to nonpaying customers would increase. Similarly, removing controls over the frequency of collections calls to delinquent customers would probably increase the number of bad debts. This section shows what would happen if various sales and receivables controls were removed and suggests possible solutions to the resulting control problems.

Documents Are Not Manually Transferred between Departments

Having the credit department manually stamp approval on an incoming customer order is a control on the granting of credit to customers. If the control were removed, orders could be shipped without the approval of the credit department, which might result in shipments to unqualified customers who will not be able to pay.

Require an Electronic Approval Code.   The computer system can freeze a customer order if it does not have an electronic approval code from the credit department attached to it. Of course, such approval codes can be stolen, so the code must be changed regularly and strictly limited to the use of the credit department staff.

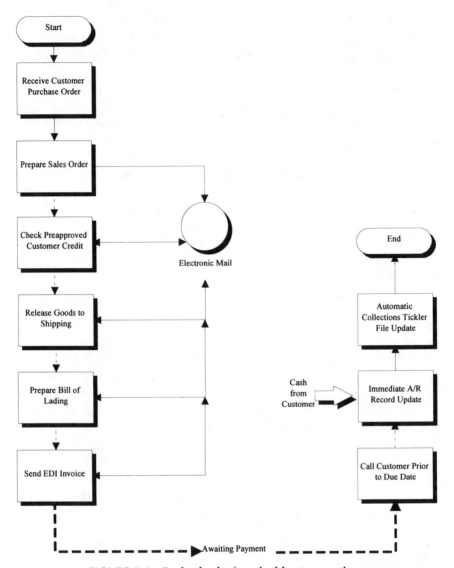

FIGURE 2.6   Revised sales/receivables transaction.

Customer Credit Is Preapproved

Reviewing a customer's credit at the time the order arrives is a control over the *current* credit situation of the customer. Preapproving customer credit means that some customer credit information may change for the worse between the time the credit is approved and the customer places the order.

Periodically Update Customer Credit Information.   The credit rating of the typical customer does not change very frequently; if bankruptcy occurs, it is usually the result of a long-term slide into insolvency that can be predicted based on a periodic review of the credit records. Consequently, a review of customer credit information two or three times a year is sufficient to cover any changes in the financial condition of the customer. If a customer requires more frequent reviews because of worsening financial condition, the company should eliminate credit for that customer anyway.

Month-End Statements to Customers Are Eliminated

Sending a monthly statement to customers is an additional means of ensuring that they will pay their bills. Eliminating the statement means that customers are receiving one less notification of the level of their indebtedness to the company, which may reduce the likelihood of payment.

Contact Customers Prior to Invoice Due Dates.   The company can replace the passive contact with the customer (month-end statements) with a more active contact that is more likely to bear fruit, such as contacting the customer prior to invoice due dates to discuss any possible problems with payment. This actually strengthens control over receivables collection, since the company's credit clerks are actively contacting customers to discover payment problems in advance rather than hoping that someone will review the periodic statement.

Invoices Are Eliminated

Sending an invoice to a customer ensures that the customer has received notification of payment and will therefore be more likely to pay. Eliminating the invoice presents a serious risk that payment to the company will not be forthcoming, since the customer has no notification of how much to pay.

Plan with the Customer in Advance.   It is extremely uncommon for customers to pay suppliers with no invoice, and many companies are not sure how to handle such payments. Because of the unusual nature of this transaction, the company and its customers must be very careful to go over all aspects of delivery and payment in advance as well as to test the new system prior to trying it out with real shipments and payments.

Record Production and Shipment Information.   In case the customer does not pay for a shipment, there must be a fallback set of information to rely on in pressing a claim for payment. This information should be a detailed record of production, backed up by shipping documents that prove the delivery was made to the customer. Thus, two other sets of information will be retained that can be used to create an invoice if this becomes necessary.

Sign a Contract.   Payments based on shipments instead of invoices are usually for long-term deliveries of parts at a set price under a blanket purchase order. If so, the company should have a signed long-term contract with each customer, so that it can be used to settle disputes regarding payment if they should arise.

Invoices Are Delivered by the Delivery Person

Segregating the invoice generation task from the shipping staff is designed to prevent the shipping department's keeping a portion of each shipment while still billing the customer for the full balance. If the invoice generation task is given to the delivery driver, it is possible for invoices to be altered.

Generate the Invoice in Advance.   By the time the delivery driver is given the invoice information, it will already have been calculated by the accounting department—after all, that is how the information got into the portable computer. Since the accounting department already has the correct shipment quantity in its records, any deviation from the original invoice will be flagged when the delivery person brings the final invoice back from the customer. The accounting department can then follow up on any significant variances. Also, these variances can be tracked over time by the accounting department to see if there are any negative variance trends for individual drivers.

Obtain the Agreement of the Customer.   Since the customer is being handed a copy of the invoice by the drier, it appears reasonable that the customer must agree that the quantity delivered is correct or else it would not agree to take the invoice. Thus, the customer acts as a control over the alteration of the invoice by being an independent reviewer of the delivered quantity.

In short, all the preceding changes result in control problems, but those problems are reduced or eliminated by the imposition of new practices that result in the same or stronger control during the transaction cycle.

These changes are minimal when compared to the long list of sales and receivables controls that are still necessary in nearly all corporate environments. The following controls should be maintained to avoid the potential increase in risk associated with their elimination.

Divide Responsibility

The majority of tasks in the sales and receivables transaction cycle must still be divided among a large number of people to avoid the risk of modification of sales and receivables information by unethical employees. This means that duties should be separated for the preparation of sales orders, credit approval, issuance of merchandise from stock, shipment, billing, approval of sales returns and allowances, and authorization of bad debt write-offs.

Use Periodic Audits

Many controls do not have to be performed continually. Instead, they are conducted on a spot basis; these are ideal tasks for the internal audit group. The following items, as well as many of the other controls listed in this section, can be assigned to the internal auditors.

Check Invoice Prices against Price Lists.   Employees may charge customers the wrong price for a variety of reasons. The audit team can compare charged prices to official prices periodically to determine if there is a continuing problem in this area. Problems most frequently arise when the pricing structure is complicated—when prices are based on volume or special deals.

Confirm Receivables Balances.   A confirmation request to a customer may reveal that an employee is not correctly applying cash received. However, a confirmation is frequently not returned by the customer, who is too busy to attend to the request.

Confirm That Credit Approval Is Required to Ship.   By various means, customer orders may bypass the credit approval process and be shipped without approval. The audit staff can examine the paperwork of company orders to discover if this is occurring.

Differentiate between Consignment and Regular Sales.   Products can be shipped to customers and still be the property of the company if there is a consignment sales agreement with the customer. This has an effect on the reported revenue number. The audit team can discover this problem by confirming sales agreements with customers and by reviewing internal sales agreements and customer purchase orders.

Look for Bill and Hold Transactions.   Products can be billed to the customer but retained at the company's location. This causes serious revenue recognition concerns, since products are supposed to be shipped before revenue is considered to be earned. The audit staff can compare billings to the shipping log to find this problem, as well as review finished goods in the company warehousing facilities.

Confirm That Invoices Are Always Issued.   If there were collusion between the shipping and accounting departments, products could be shipped to a false address without an invoice being generated, resulting in employee ownership of the product. This can also occur if products are shipped without an entry in the shipping log, which does not require collusion. The audit team can look for this problem by examining inventory variances, comparing bills of lading to the shipping log, and comparing bills of lading or the shipping log to invoices issued.

Compare Bad Debt Write-Offs to Cash Received.   An employee with control over bad debt write-offs and cash receipts can write off accounts that are collectible and then pocket the money when it arrives. The audit team can find this problem by comparing bad debt write-off amounts to the cash receipts record maintained by the mail room or by reviewing the receipts record related to the company lockbox.

Confirm That Invoices Are Issued in the Correct Amount.   An employee with control over both the invoicing and cash receipts functions could invoice a customer more than the correct amount, record the correct amount in the general ledger, and pocket the excess amount of cash when it arrives. The audit team can find this problem by reviewing canceled checks to ensure that they are being cashed into the company's account, comparing individual invoice copies to the sales journal, and independently calculating the correct amount of invoices based on the shipping log and comparing this to the amounts actually invoiced.

Reconcile Detail to General Ledger

In many companies, the detail balance does *not* match the general ledger balance more often than it *does* match. One ploy that may arise in this area is when an employee falsely increases the general ledger receivables balance or issues false invoices in order to present a large borrowing base to the company's bank, which then issues a loan to the company based on the size of the borrowing base. A monthly reconciliation of the two amounts may reveal that incorrect or inappropriate changes to either balance are being made.

Require Approval of Discounts, Returns, and Allowances

It is possible for an employee to issue an unauthorized discount to a customer in exchange for a kickback. Alternatively, this may be a way to collect the amount of the discount when payment is received from the customer (though this requires collusion with the person recording cash receipts or the combination of the two functions with one person). This problem is avoided if supervisory approval is required in order to process a discount, return, or allowance.

Require Some Review of Customer Credit

Despite the suggestion earlier in this chapter for preapproving customer credit, it is still necessary to review customer credit in some manner at some point before the product is shipped to the customer. At a minimum, the computer system must compare the amount of a new order to the amount of remaining credit assigned to each customer and flag those orders that exceed their designated credit levels.

Review the Period-End Cut-off

The company may report an incorrect amount of revenue in a period if items are billed that have not been yet shipped. The company can avoid this problem by comparing invoicing records to the shipping log at the end of each reporting period and moving invoices into the next reporting period if they have been issued prematurely. This procedure also discovers items that have been shipped but not billed.

In summary, most controls related to the sales and receivables cycle cannot be changed without seriously weakening a company's control structure. However, using technology or alternative processing steps can reduce the number of control steps used in a typical transaction.

QUALITY ISSUES

The introduction of mistakes into the accounting process can slow it down considerably because staff must then take the time to track down the mistake, correct it, discern the reason for the error, and fix that problem as well. Thus, high-quality sales and receivables processing is defined as including the fewest mistakes. This section lists the most common sales and receivables processing mistakes and suggests how to prevent them.

When accounting transactions are sent electronically to an employee for approval or review, it is possible for the transactions to be delayed if the employee does not access electronic mail for a variety of reasons—sickness, vacation, or termination. When this happens, the transaction cannot move further along in the chain of transaction steps. To avoid the problem, some electronic mail systems will notify the sender of the electronic mail message if the message has not been accessed within a specific time period. Another backup method is to have the electronic mail routed to both a primary recipient and a secondary recipient, so that the likelihood of action being taken is doubled.

When computerized accounting systems are used throughout the company to process shipment information, create invoices, and record cash receipts against receivables records, there is a risk that the entire staff will not be able to function if the hardware fails. This problem can be avoided by using uninterruptible power supplies, power surge suppressers, daily backups of the entire system, multiple tape backups, off-site tape backup storage, and mirrored hard drive storage.

When invoices are eliminated, the company can be harmed financially if it does not keep perfect records of shipments: it would have to depend on customers' records to receive correct payment for shipments received. In addition, if a customer does not pay for any reason, the company has no way to prove its claim regarding the amount of product shipped. To solve this problem, the company must maintain a shipping log. To further ensure that the shipping records are kept up, the company's internal audit team should periodically compare the company's production records to the shipping log to look for any major discrepancies.

The most common error in the sales and receivables transaction cycle is to not invoice the customer. This happens when there is a weak link between the shipping department and the accounting department; shipping information does not necessarily reach the accounting staff. This can occur when the shipping log is not sent to the accounting staff at all (forgetfulness on the part of the shipping staff), when the shipping information is lost in transit between the shipping department and the accounting department, and when the accounting department neglects to create and mail the invoice. The necessary link between the shipping and accounting departments can be eliminated by automation—the shipping staff types shipment information directly into a terminal, and the information is automatically conveyed to the accounting department by the computer system. If there is a projected ship date on a customer order in the computer system, a report can be created that lists the orders in the system with overdue ship dates or no invoices (see Table 2.3). This report flags orders that may have been shipped but not invoiced. The controller can print out such a report once a week and investigate any possible problems.

Another problem that may arise in the link between the shipping and accounting departments is when the shipping staff does not enter the shipment into the shipping log at all. The best way to avoid this is to tie the creation of the bill of lading to the automatic creation by the computer system of an entry in the shipping log. Since the freight company will not accept a shipment without a bill of lading, this ensures an entry in the shipping log.

Another error is the shipment of products to customers whose requests for credit have been turned down. This is caused by a breakdown in procedures,

TABLE 2.3   Orders in the System with No Invoices

Customer	Order No.	Amount ($)	Projected Ship Date
Able Phone Systems	0001123	14,321	10/10/96
Bosch Tissues Inc.	0001127	9,267	10/08/96
Cummins Lighting Co.	0001097	8,401	09/30/96
Dorfman Thermometers	0001092	3,232	08/31/96
Englehart Shipping Lines	0001199	9,461	10/03/96
Finley Truck Lines	0001204	1,001	10/02/96
Grimm Kid's Games, Inc.	0001207	2,003	09/22/96

because no orders should be shipped until credit approval has been given. One way to fix this problem is approving credit in advance. Another method is to enforce the credit/shipping procedure by regular reviews of the system by the internal audit department. Also the system can be automated so that the computer will not allow a customer order to be listed on the production schedule unless a credit approval code has been entered for that order.

When the customer sends in a purchase order, it is possible for the purchase order to be lost by the company before it even enters the sales/receivables transaction cycle, resulting in a lost sale or a very dissatisfied customer. An order can arrive in the mail, over the phone, at the fax machine, or by hand. The order can be lost at any of these points of arrival. To solve the problem, the fax machine can be set up in the order entry department, thereby reducing the chance of an order's being lost by someone outside of the department. The fax machine can also have a considerable amount of memory, so that documents are stored in the fax machine's memory even if there is no paper in the machine. If an order is delivered by phone, the reception staff must route the call to a central order entry phone number that has a voicemail backup, so the order can be stored in voicemail even if no one is there to take the call. If the order is delivered by mail, the company can install a scanner in the mail room, so that purchase orders are entered into the system prior to delivery of the paper document to the order entry department. This means that a digital record is kept of the purchase order, even if the paper record is lost. Finally, if an order is delivered by hand, it should be routed through the mail room first, so that the document can be scanned (the same procedure as used for documents delivered by mail) prior to being delivered to the order entry department through the interoffice mail system. In short, the risk of losing purchase orders can be mitigated by installing backup systems that keep a record of the order.

In summary, most of the errors common to the sales/receivables transaction cycle can be counteracted by various fail-safe measures that back up the primary transaction, so that a second person or the previous person involved in the transaction is warned if further action is not taken. Other measures involve eliminating the faulty transaction step through automation. When these measures are implemented, the incidence of errors drops dramatically.

COST/BENEFIT ANALYSIS

This section demonstrates cost/benefit analysis for electronic mail, preapproval of customer credit, the elimination of month-end statements, an automated collections tickler file, electronic data interchange, calls prior to invoice due dates, and the elimination of all invoices. In these examples, expected revenues are as realistic as possible.

Use Electronic Mail to Move Transactions

The management of TechnoStuff, a distributor of electronic supplies, wants to increase the speed of accounting transactions by using electronic mail to move transactions between employees. The company's MIS director, however, claims that this would require a "monstrous" amount of programming time and major changes in the company's packaged accounting software, which would have to be made every time the company purchased an upgrade to the accounting software. Further investigation reveals that for $150,000 an accounting software package can be purchased that incorporates electronic mail approvals into its accounting functions. The data conversion, implementation, and training costs equal the purchase cost. The savings to be gained relate primarily to the competitive advantage secured by shrinking the order-processing time from three days to 4 hours. The sales department estimates that it can secure a sales increase of 5% by advertising the high-speed ordering system. The company's annual sales are $23,000,000, and its gross margin is 30%. Should transactions be conducted by e-mail?

Solution.   The cost of installing e-mail transactions is the cost of purchasing and implementing a new accounting software package that incorporates the e-mail feature. The savings relate to sales gained from the competitive advantage of having a faster order-processing time.

Cost of Installing E-Mail

Purchase cost of software	$150,000
Data conversion, implementation, and training costs	150,000
Total software cost	$300,000

Benefit of E-Mail

Company sales volume/yr	$23,000,000
Increase in volume resulting from e-mail	× 5%
Increase in volume/yr	$ 1,150,000
Gross margin on sales	× 30%
Incremental margin on sales increase	$ 345,000

With costs of $300,000 and benefits of $345,000, the cost/benefit analysis indicates that the software should be purchased.[2]

Preapprove Customer Credit

Adi Pinkerson, the CFO of the CD Warehouse distribution center, is concerned about the time spent qualifying CD retail stores for credit after orders have been placed. She notes that many of the company salespeople are disappointed when they pursue potential customers and close deals, only to have their credit turned down by the credit department. She decides to look into switching over to pre-approved credit before the salespeople make any sales calls. The sales department estimates that 2 hours of the sales manager's time will be needed each month to collect lists of potential new customers from the sales staff and pass along that information to the credit department. The sales manager earns $85,000 per year ($40.87 per hour). The list will contain an average of 40 new customers. The collections staff must take an hour to collect and review credit information about each potential customer. The average hourly pay for each collections clerk is $8.90. Ms. Pinkerson estimates that this advance work will keep the credit department from turning down an average of five customers per month while saving the sales staff an average of 8 hours to secure each of the five orders. The average sales employee earns $45,000 per year ($21.63 per hour). Should customer credit be preapproved?

Solution.   The savings come from having fewer credit checks to turn down and a significant decrease in the work of the sales department, while the additional cost primarily comes from having to prequalify a large number of potential customers who may not even place an order with the company.

---

[2] This is a difficult cost justification for many companies, since the software cost is quite high. When estimating sales increases based on the competitive advantage of having faster order-processing speeds, considerable caution (if not skepticism) should be used. A survey of customers, asking them how a faster order turn-around time will affect the size and frequency of their orders, is helpful in making this determination.

Cost of Preapproving Customer Credit

Sales manager time/month to collect information	2 hrs
Months/yr	× 12
Hours/yr	24 hrs
Sales manager cost/hour	× $40.87
Total sales manager cost/yr	$ 981
Credit clerks time/month to review customer credit	40 hrs
Months/yr	× 12
Total hour/yr	480 hrs
Credit clerk cost/hour	× $8.90
Total credit clerk cost/yr	$ 4,272
Total cost/yr of preapproving customer credit	$ 5,253

Benefit of Preapproving Customer Credit

Time/month saved by avoiding credit reviews	5 hrs
Months/yr	× 12
Hours of credit reviews avoided	60 hrs
Credit clerk cost/hour	× $8.90
Total savings/yr from avoiding credit reviews	$ 534
Customers turned down/month for bad credit	5
Months/yr	× 12
Customers turned down/yr	60
Salesperson time to secure an order	× 8 hrs
Salesperson time/yr to secure bad credit orders	480 hrs
Salesperson cost/hour	× $21.63
Total cost/yr of unnecessary sales calls	$10,382
Total savings/yr preapproving customer credit	$10,916

With costs of $5,253 and savings of $10,916, it is evident that the change should be made. The key reason for the change (in terms of cost/benefit) is that it saves the time of the sales department, which is more expensive per hour than the cost of the collections department.[3]

Eliminate Month-End Statements

Ms. Bentley, the controller of Ancient Autos Warehousing, is concerned that month-end statements are not generating any savings for the company, which

---

[3] The reduction in the work flow caused by removing the credit check does not even enter into the cost/benefit calculation, since the largest cost savings do not relate to a quicker order-processing function. However, a cost/benefit analysis could include the competitive advantage offered by having a fast order turn-around time.

sends out 1,250 statements per month to its customers. After a three-month test period when no month-end statements were sent, the company's days receivables measure increased from 50 days to 51 days, an increase of 2%. The company charges each customer an average of $200 per month to store his expensive car. The company earns 9% on its investments. Ms. Bentley spends $250 per month on statement forms and another $0.75 in labor and mailing costs to send out each statement. Should the company eliminate month-end statements?

Solution. The cost savings come from the materials, postage, and labor costs associated with a period-end statement mailing. These savings are offset by a slight worsening of the company's receivables balance.

Cost of Deleting Period-End Statement Mailing

No. of customers	1,250
Average balance/customer	× $200
Total average receivables balance	$250,000
Increase caused by lack of statements	× 2%
Amount of increase	$ 5,000
Investment rate of interest	× 9%
Interest cost of excess receivables	$ 450

Benefit of Deleting Period-End Statement Mailing

Cost/month of statement forms	$ 250
Months/yr	× 12
Cost/yr of statement forms	$ 3,000
No. of customers	1,250
Months/yr	× 12
Invoices mailed/yr	15,000
Mailing cost/statement	× $.75
Statement mailing cost/yr	$ 11,250
Total savings/yr from deleting mailing	$ 14,250

The cost of eliminating monthly customer statements is $450, versus printing and mailing savings of $14,250. This cost/benefit analysis indicates that customer statements should be eliminated.[4]

Deliver Invoice with Shipment

John Walker, president of Walkers Deluxe, manufacturers of top-quality hiking shoes, wants to reduce the time needed to invoice customers. One possibility is to

---

[4] Efficiencies associated with shifting accounting staff to other functions are more difficult to document but could be added to this analysis.

have the company's delivery drivers carry portable computers that are loaded with invoicing information. The CFO, David Slog, reviews the costs and benefits of installing such a system. Each of the company's ten delivery drivers would be equipped with a portable computer costing $2,200 and a $300 printer. In addition, each of the drivers would need an hour of training. The average driver wage is $16 per hour. By delivering the invoices at the same time as the products, Mr. Slog estimates, invoices would be paid two days sooner than usual. Since the current days' receivables statistic for the company is 41 days, this would be a drop of 4.9%. Walkers Deluxe has an average outstanding receivables balance of $3,450,000 and earns 8.5% on its investments. Should this system be adopted?

Solution.   The cost savings are from reducing the company's receivables balance. These savings are primarily offset by the cost of computers and printers.

Cost of Delivering Invoice with Shipment

Cost of portable computer	$ 2,200
Cost of printer	300
Computer hardware cost	$ 2,500
No. of drivers	× 10
Total computer hardware cost	$25,000
No. of drivers	10
Training time required/driver	× 1 hr
Total training time required	10 hrs
Driver pay/hour	× $16
Total cost of training	$ 160
Total cost of delivering invoice with shipment	$25,160

Benefit of Delivering Invoice with Shipment

Current receivables balance	$3,450,000
Reduction in receivables with new system	× 4.9%
Amount of receivables reduction	$ 169,050
Corporate return on investments	× 8.5%
Savings from delivering invoice with shipment	$ 14,369

The cost of delivering invoices with the shipment is $25,160, which is counteracted by savings of $14,369. With a payback of 1.75 years, it appears reasonable to proceed with the project.[5]

---

[5] When the cost/benefit analysis is based on an estimate of receivables reduction, it is best to try a pilot test with a small number of customers to see if the receivables decrease is really going to happen. If not, then the controller will have avoided the much greater expense of a full implementation without sufficient payback.

Install Automated Collections Tickler File

Dudley Anderson, the collections supervisor at Smith & Smith Gunsmiths, goes to a collections conference and hears about automated collections tickler systems. He wants one. As the controller, you review the marketplace and find that such a system must be designed, not bought, since the only way to purchase one is as part of a complete accounting package. It is too expensive to switch accounting packages just for the automated tickler file. To build one in-house, the programming staff requires 2,200 hours for design, programming, testing, and documentation. The average programmer earns $22 per hour. In addition, a total of 8 training hours will be needed for a collections clerk, plus 20 hours more for her to meet with the programming staff initially and go over requirements for the new system. Once the system is installed, Mr. Anderson believes, the collections staff will virtually eliminate its record-keeping time, which will allow him to cut one staff person (who earns $16,500 per year, or $7.93 per hour) while still increasing the number of collections calls by 30%. The increase in calls should reduce the company's days of receivables from 47 to 42, which is a 10.6% decrease. The company's average total receivables balance is $4,525,000, and its borrowing rate of interest is 8%. Should the collections supervisor get his system?

Solution. The cost savings come from reducing the receivables balance as well as from reducing the number of staff members needed in the collections department.

Cost of Installing Tickler System

Programming time required	2,200 hrs
Programming cost/hour	× $22
Total programming cost	$48,400
Collections time required	28 hrs
Collections clerk cost/hour	× $7.93
Total collections cost	$ 222
Total cost of creating tickler system	$48,622

Benefit of Installing Tickler System

Eliminate one clerk position	$ 16,000
Receivables balance	$4,525,000
Balance decrease with tickler system	× 10.6%
Decrease in receivables balance	$ 479,650
Borrowing rate of interest	× 8%
Interest earned on reduced capital needs	$ 38,372
Total savings after installation	$ 54,372

With a cost of $48,622 and savings of $54,372 in its first year of operation, the cost/benefit analysis indicates, the automated collections tickler system should be implemented.

Use Electronic Data Interchange

Ms. Stuart, the founder of Old-Tyme Furniture, Inc., wants to electronically bill her customers. The company has 34 large customers, and another 122 mom-and-pop customers with no computerization. The average invoice is for $1,000 and requires 5 minutes to generate manually. The fully burdened cost of an invoicing clerk is $13 per hour. The cost of purchasing EDI software is $2,500. A value-added network (VAN) would transfer the transactions to the 34 largest customers. The charge per transaction by the VAN is $0.50. The average customer is sent 17 invoices per year. The MIS director estimates that programming the interface between the EDI software and the company's accounting software will take three months of a programmer's time. The fully burdened payroll cost of the programmer is $38,000. The average mailed invoice is received in 45 days, and invoices will be received in 33 days if EDI is used, since there will be no missing receivables transactions to track down. The company's total average receivables amount outstanding per month is $492,000 from its 34 largest customers, and the interest rate on its bank debt is 9%. With the EDI system in place, the work load of the collections clerk will be reduced by one-third. The collections clerk earns $22,000 per year. Should EDI be used?

Solution.   The cost savings are from eliminating the labor associated with the invoices as well as from reduced interest costs because of quicker collection of receivables and reduced collection costs.

Cost of EDI Implementation

Cost/yr programmer	$38,000
Three months' work	× .25
Total programming cost	$ 9,500
Cost of EDI software	2,500
Total EDI implementation cost	$12,000

Cost of EDI Transactions

No. of customers with EDI	34
Annual transactions/customer/yr	× 17
EDI transactions/yr	578
Cost/transaction	× $.50
Total transactions cost/yr	$289

Labor Savings With EDI

No. of customers with EDI	34
Invoices/customer/yr	× 17
Invoices/yr	578
Time to create invoice	× .083 hr
Total time to create invoices	48 hrs
Clerical cost/hour	× $13
Total savings from no invoices	$ 624
Cost of collections clerk/yr	$22,000
Time savings with EDI	× .33
Total savings with EDI	$ 7,260

Interest Savings With EDI

Average receivables days outstanding	45
Average days outstanding with EDI	− 33
No. of days reduction in A/R	12
Average receivables balance	$492,000
Percentage reduction in A/R balance	× 27%
Reduction in receivables balance	$132,840
Interest rate on bank debt	× 9%
Total savings from reduced receivables	$ 11,956

Since the total cost is $12,289 and the total benefit is $19,840 for one year of operation, the cost/benefit analysis indicates that an EDI system should be implemented.[6]

Call Customers Prior to Invoice Due Dates

As the CFO of Fundamental Dynamics, Inc., you are concerned that many of your company's invoices are lost in the midst of your customers' approval processes. This problem is aggravated by the large size of the invoices for the capital equipment that the company sells. As a result, the collections staff is constantly faxing invoices to customers or sending invoice copies by overnight mail. The collections manager estimates that one-third of all customers lose their invoices (either temporarily or permanently) in the approval process and require new invoice copies. This results in a drastic lengthening of the days of receivables, from 42 days to 54 days (a 29% increase). The company's average total

---

[6] If the analysis is carried forward several years, the only continuing cost is for each VAN transaction, whereas the benefits continue to pile up. Thus, a multiyear analysis will yield even more convincing results.

receivables balance is $3,150,000. The interest rate on the company's debt is 10%. The collections manager estimates that it will require an extra collections clerk to contact all customers in advance of their invoice due dates in order to figure out which one-third of the total customer base has lost its invoices. The collections clerk will cost $20,800. The company sends 129 invoice copies per year to customers by overnight mail. The per-piece overnight mail cost is $7. Should prior calls be implemented?

Solution. The cost savings are from reducing the days of receivables outstanding and reducing overnight mail costs, whereas the labor cost associated with the extra phone calls offsets the savings.

Cost of Calling Customers Prior to Invoice Due Dates	
Cost of collections clerk	$20,800
Benefit of Calling Customers Prior to Invoice Due Dates	
Average receivables balance	$3,150,000
Percent of balance due to lost invoices	× 29%
Amount of receivables due to lost invoices	$ 913,500
Interest rate on company debt	× 10%
Interest cost due to lost invoices	$ 91,350
No. of invoices sent by overnight mail	129
Cost/overnight mail package	× $7
Total cost of overnight mail	$ 903
Total savings from early customer contacts	$ 92,253

Since the total cost is $20,800 and the savings are $92,253, the cost/benefit analysis indicates that customers should be called in advance.[7]

Eliminate All Invoices

Mike Barrie, owner of Pan Pizza Equipment, Inc., has been contacted by several customers who would like to start paying the company based on their production. Pan Pizza produces the rollers that are used to move partially cooked pizzas through a baking oven. The customers will build the rollers into their baking

---

[7] The proportion of invoices lost in this example is accurate for capital equipment, because so many signatures are required for large invoices, resulting in invoices being lost as they are routed to collect signatures. However, an analysis for smaller invoices would yield a much lower percentage of lost invoices.

machines, which require one week to build, and then pay Pan Pizza within two additional weeks. This means that payment will occur in three weeks instead of the usual 30 days (a reduction of 30%). The customers proposing this change make up 65% of Pan Pizza's total sales volume. Its average receivables balance is $1,560,000. The controller feels that there will be difficulty matching customer payments to any internal documents for control purposes and proposes having some customized programming work done that will match payments received to the company's cumulative record of materials sent to each customer. The programming staff estimates that this project will require four months of programmer time (at an average salary of $38,000). The controller is also concerned that, without invoices, the company may have some difficulty proving to the auditors that it has booked the correct revenue each year. The controller decides to bring in the auditors for a review of the proposed transactions; this consulting work will cost $4,000. Also, the customized software will not automatically link to the company's packaged accounting software, so a one-day reconciliation must be manually performed each month by the assistant controller between the results of the customized software and the standard software. The assistant controller earns $48,000. Pan Pizza earns 8% on its investments. Should this new system be implemented?

Solution. The proposal has costs related to tracking the incoming payments and matching those payments to the company's sales. These costs are offset by greatly reduced accounts receivable balances.

Cost of Eliminating Invoices

Cost/yr of programmer	$38,000
Four months' work	÷ 3
Cost of programmer	$12,667
Cost of auditor review	$ 4,000
Cost/yr of assistant controller	$48,000
No. of business days/yr	÷ 280
Cost/day assistant controller	$171.43
No. of reconciliation days/yr	× 12
Cost of reconciliation	$ 2,057
Total cost of eliminating invoices	$18,724

Benefit of Eliminating Invoices

Average receivables balance	$1,560,000
Percentage reduction in receivables	× 30%
Dollar reduction in receivables	$ 468,000
Interest rate	× 8%
Total savings from reduced receivables	$ 37,440

With costs of $18,724 and savings of $37,440, the cost/benefit analysis indicates that the company should adopt a no-invoice system.[8]

In summary, cost/benefit analyses can prove that the changes advocated in this chapter will be worthwhile to the company. However, the analyses are usually founded upon cost savings that do not relate to the speed of the transaction, since transaction speed is difficult to quantify. Thus, the controller must frequently rely on other "incidental" savings when trying to justify changes whose primary objective is to streamline the accounting operation.

REPORTS

The changes discussed in this chapter require several new reports. For example, a preapproved customer list is needed as well as collections tickler file reports, lists of unacknowledged EDI invoice transmissions, and lists of customers to call about upcoming invoice due dates. All these reports contribute to the efficient, high-speed functioning of the accounting department.

A preapproved customer list that includes the amount of total credit and available credit is useful for management review, since it highlights those customers with low credit and whose credit has been almost fully utilized. It should not be used each day by the staff to check against incoming purchase orders, since the report may become outdated quickly. It is better to have the order entry staff use on-line credit information in order to have the most up-to-date data. A preapproved customer credit list is shown in Table 2.4.

An appropriate use of computerization is the automation of the daily call list for the collections staff. The computer system can automatically update the overdue accounts list as money is received by the company, and also add overdue accounts to the list as they become overdue. An automated collections report is shown in Table 2.5.

The controller needs to know about any EDI invoice transmissions for which acknowledgments have not been received. This information allows the company to contact the customer to verify that the invoice was received, so that it will be paid on time. An EDI invoice acknowledgment report is shown in Table 2.6.

---

[8] The analysis does not include the effect of reducing the amount of invoicing work, because many companies do not have accounting software that can handle payments without invoices. If a company were able to account for such payments without having to generate invoices, it could also reduce the labor needed to produce invoices. Also, many customers will not give preferential payment terms when switching over to a no-invoice system. In that case, the company can only save money by reducing the labor of the invoicing clerk, which may require considerable programming time to accomplish. Thus, the payback period is greatly lengthened.

TABLE 2.4 Preapproved Customer Credit List

Customer	Total Credit Granted ($)	Total Credit Used ($)	Unused Credit ($)	Percentage of Unused Credit	Amount of Overdue Invoices ($)
Albatross Mining	40,000	20,000	20,000	50%	2,000
Brass Refinishing	10,000	8,000	2,000	20	4,000
Cards R Us	5,000	0	5,000	100	0
Davidson Metals	12,000	1,000	11,000	92	0
End Run Printing	15,000	14,000	1,000	7	10,000
Gore Minerals	62,000	60,000	2,000	3	50,000
Highway Counters	33,000	13,000	20,000	61	4,500

TABLE 2.5 Call List for Collections

Customer	Phone No.	Contact Name	Invoice No.	Amt Due ($)	Comments
Englehart Co.	719-112-6784	Evelyn	6742	567.21	Faxed invoice
Fop & Mop Corp.	508-231-4795	George	6751	330.33	Claims goods damaged
Grady's Ltd.	303-312-5678	Frank	6702	7,555.00	Needs shipping trace
Honcho & Son	202-331-0932	Bill	6631	921.12	Claims goods damaged
Jobson & Clark	212-113-9074	Martha	6522	888.77	Faxed invoice
Killiwary Birds	308-222-3322	Mabel	6780	469.42	Paying on installment
Lanterns Plus	415-333-1111	Liz	6411	902.01	Refuses to pay

TABLE 2.6 Unacknowledged EDI Invoices

Customer	EDI Transaction No.	Invoice No.	Invoice Amt ($)	Date of Transmission
Englehart Co.	1001	567231	1,213.00	5/12/98
Fop & Mop Corp.	1003	567233	542.13	5/13/98
Grady's Ltd.	0999	567229	600.11	5/11/98
Honcho & Son	1000	567230	9,045.14	5/12/98
Jobson & Clark	1006	567236	13.11	5/14/98
Killiwary Birds	1008	567238	154.98	5/15/98
Lanterns Plus	0998	567228	2,333.00	5/10/98

TABLE 2.7   Call List for Upcoming Invoice Due Dates

Customer	Contact Person	Phone No.	Invoice No.	Invoice Amt ($)	Invoice Due Date
Englehart Co.	Evelyn Gregson	719-112-6784	6742	567.21	5/31/98
Fop & Mop Corp.	George Anders	508-231-4795	6751	330.33	5/30/98
Grady's Ltd.	Frank Horton	303-312-5678	6702	7,555.00	5/27/98
Honcho & Son	Bill Matthews	202-331-0932	6631	921.12	5/28/98
Jobson & Clark	Martha Davis	212-113-9074	6522	888.77	5/29/98
Killiwary Birds	Mabel Smith	308-222-3322	6780	469.42	5/31/98
Lanterns Plus	Liz Jones	415-333-1111	6411	902.01	5/30/98

A good way to avoid overdue invoice payments is to call customers in advance to ensure that the invoice was received and that it has been approved and is ready for payment. If either of these has not occurred, then the collections department can work on getting the proper information to the customer before the due date. A call report for upcoming invoices due is shown in Table 2.7.

In summary, the sample reports shown in this section support bypassing steps in the normal sales transaction, reducing the work required to complete existing steps, and spotting problems as they occur, so that they can be fixed without delay. When used as part of the overall improvements mentioned in this chapter, these reports contribute to the increased speed of the sales and receivables transaction cycle.

TECHNOLOGY ISSUES

This section examines the ways in which recent advances in technology can speed up the sales and receivables transaction cycle. These advances include the transfer of transactions between employees by electronic mail and the delivery of invoice information to the customer on a diskette or portable computer. The first method reduces the amount of time required to move transaction-related paperwork between employees, and the second reduces conflicts with customers over the content of deliveries, which speeds payment.

Paperwork related to sales transactions can be passed from employee to employee by electronic mail. This avoids problems with paper-based transactions being lost when they are moved from person to person. For example, a typical customer order moves from the mail room to an order entry clerk, to a credit clerk, to the production scheduling department. If the paperwork is lost during any of these transfers, the order will not be processed, and the company will lose revenue. To avoid this problem, the order entry clerk enters the order into the accounting database, which safely stores the record in the centralized data storage location so that it cannot be lost. Once this is done, the software searches the order record to see if the customer requires approval from the credit department. If so, the software sends an electronic message to the credit department, asking for a review of the order. The credit clerk then accesses the record directly from the electronic message and enters a credit approval code in the record, which is still stored in the accounting database. The key point is that the order information cannot be lost; it stays in one place while the software sends messages out to the employees via the company's computer network, requesting updates to the record. Most computerized accounting packages do not yet include this feature, so considerable customization may be required to include it in a company's existing software.

Invoice information can be given to customers on diskettes or portable computers. The delivery person carries a hand-held device that can print the invoice on the spot or a diskette containing the delivery information. In either case, the data include part numbers and descriptions, quantities, and prices. The customer's receiving personnel can count the incoming parts or accept the shipment without review, depending on whether there has been a history of accurate shipments. If there are count differences, the delivery person inputs the change into the portable computer, prints out a revised invoice, and gives it to the receiving staff. The delivery person then delivers the diskette back to the company's accounting department, which processes the revised invoicing information but does not have to send an invoice to the customer. This eliminates the time otherwise needed to bring the shipment information back to the company, transfer it to the accounting department, create and print an invoice, and mail it back to the customer. Also, if the diskette already contains all the information listed on an invoice except the price, and the price has been negotiated in advance, the need for an invoice has been eliminated. The customer does not need a printed invoice, just the delivered quantity information. In either case, this option is usually only available if the company delivers products via its own staff instead of through a third-party carrier.

In summary, the use of technology can speed up the sales and receivables transaction cycle by shrinking move and wait times and by eliminating the time needed to create and mail invoices to customers.

MEASURING THE SYSTEM

The suggestions for improvement in this chapter relate to improving the speed of the sales/receivables transaction cycle and reducing the transaction processing work of the accounting staff.

Measurements to Track Speed of Sales/Receivables Transactions

Time from Receipt of Order to Production Scheduling.   This covers the time period in the sales transaction that applies to processing the incoming customer order. The best ways to improve this measure are to eliminate any move and wait times in the process and to bypass or automate as many other steps as possible. A final option is to perform functions in advance of the transaction (such as prequalifying customer credit).

Number of Days' Sales Outstanding.   This covers the time period between when the product has been shipped but before the payment has been received, and tracks the speed of collection. The best ways to improve this measure are to rigorously reject the orders of those customers with bad credit records, review invoice approvals with customers prior to payment due dates, and accelerate the transmission of invoices to customers with overnight delivery services (if the size of the invoice warrants the cost).

Time from Receipt of Cash to Updating of Receivables Records.   This covers the final phase of the sales and receivables transaction cycle. The best ways to improve this measure are to adopt an accelerated delivery of checks from the mail room to the accounting department, and to use a computerized accounting system to offset cash receipts against receivables records.

Number of Value-Added Steps in the Transaction Cycle as a Percentage of the Total Number of Steps.   This covers the entire transaction cycle and is an excellent measure of the efficiency of the entire process. Only about 10% of the steps in a typical transaction cycle are value-added. If the number of steps can be streamlined so that the ratio exceeds 50%, the process is extremely efficient. A ratio of 80% or better is world-class.

Of the preceding measurements, only number of days' sales outstanding has a commonly accepted formula:

$$\text{Average no. of days' sales outstanding} = \frac{\text{Average receivables}}{\text{Annual sales on credit}} \times 365$$

For example,

$$\frac{\text{Average receivables}}{\text{Annual sales on credit}} = \frac{\$204,510,000/12}{\$122,220,000 - \$98,400} \times 365 = 51 \text{ days' sales outstanding}$$

A well-managed collections operation should maintain a days' sales outstanding figure that is about one-third beyond the terms of sale. For example, if invoices are due in 30 days, an acceptable days' sales outstanding figure would be 40 days. If the days' sales outstanding appears to be abnormally long, the controller should verify that only receivables from commercial accounts (not receivables for officer notes or taxes) are included in the receivables total. Removing noncommercial receivables from the total will improve the ratio. Another underlying problem causing a long days' receivables ratio is giving especially long payment terms to customers. Special payment terms, paradoxically, are commonly given to the worst customers, who cannot afford to pay under the company's normal payment terms. Eliminating these special deals will improve the ratio and probably reduce the need for a bad debt reserve at the same time.

Measurements to Track Reduction of Work of Accounting Staff

The controller should formulate and compile these costs periodically on a project basis to see if any changes in the costs are occurring. These costs are not ratios with specific formulas; the detail costs that make up each measure will vary by industry. However, the following can be used as a starting point for most businesses.

Cost to Process an Incoming Order.   Divide these costs by the number of incoming orders to derive the cost per order:

- Mail room labor to open mail and forward purchase orders.
- Labor to create a sales order based on the purchase order.
- Labor to conduct a credit check of the customer.
- Phone cost of contacting the customer about credit information.
- Labor cost of entering the order in the production system.
- Labor for filing of paperwork.
- Cost of physical storage.

Cost per Invoice Issued.   Divide these costs by the number of invoices issued to derive the cost per order:

- Labor to record shipping information in the shipping log.
- Labor to create an invoice based on the shipping information.
- Cost of the invoice paper and mailing materials.

- Cost of postage.
- Cost of mail room labor to mail the invoice.

Cost of Cash Application.   Divide these costs by the number of payments received to derive the cost per receipt.

- Mail room labor to open mail and forward checks.
- Labor to apply cash against receivables records.
- Labor to contact customers about application problems.
- Phone cost to contact customers.
- Labor to file invoice and payment information.
- Cost of physical storage.

Cost of Collection per Invoice.   Divide these costs by the total number of invoices outstanding to derive the cost per invoice.

- Labor to contact customers.
- Cost of phone calls to customers.
- Cost of overnight or fax transfer of information to customers.
- Labor for document filing.
- Cost of physical storage.

The cost per invoice issued, according to a recent study,[9] varies widely by type of industry and size of firm. For example, the cost to issue an invoice in the service industry is $0.33, whereas it more than doubles, to $0.78, in the manufacturing area. However, those costs only relate to the isolated case of creating an invoice. When the entire sales and receivables transaction cycle is taken into account, the benchmark rises to $4.60, versus $15.00 for the Fortune 100 companies.[10]

In summary, most of the performance measures advocated in this section are not the standard ones included in most accounting systems; the controller must decide if it is worthwhile to assign staff to track these new measurements. A compromise solution is to track these measurements on a sample basis periodically to determine if performance is changing appreciably over time. This may be a task for the internal audit department.

---

[9] Hackett Group, *AICPA/THG Benchmark Study: Results Update and Analysis,* 1994.

[10] Steve Coburn, Hugh Grove, and Cynthia Fukami, "Benchmarking with ABCM," *Management Accounting,* January 1995, p. 59.

IMPLEMENTATION CONSIDERATIONS

A number of problems relating to change management may arise when revised sales and receivables systems are installed (see Chapter 11). The following problems can be expected when implementing revised sales and receivables systems.

Electronic Mail

Implementing the movement of accounting transactions with electronic mail is extremely difficult, because accounting transactions must be taken out of the accounting system, sent to user mailboxes, modified, and sent back to the accounting system. Since one of the primary elements of an accounting system is security, most packages do not allow this. There are two ways to avoid the problem. First, customized programming can alter the software code. This option is very expensive and time-consuming, and also means that if the accounting software is purchased, then updates to the software may destroy any changes made to the software. In short, this is a difficult option to implement. The better option is to use one of the newer accounting packages that incorporates the use of electronic mail into its operations. Even when such a package is purchased, however, the company must still endure a (usually) painful and (always) expensive conversion from its current accounting system to the new system. Because of the expense related to the transition, the controller will usually need to find additional reasons for switching software than just the addition of transaction transfers by electronic mail.

Preapproval of Customer Credit

The problem most frequently encountered when requiring preapproval of customer credit is persuading salespeople to work with the credit department in advance. Many salespeople resent having too many procedures forced upon them when they can be out in the field selling and earning commissions. The best way to sell them on this change is to point out the reduction in processing time that occurs when credit approval is taken out of the sales transaction. Since the salespeople may be skeptical about this, it may be worthwhile to implement a pilot project with the most willing salesperson and track the actual time reduction for the transactions of that salesperson's customers. Once this information is made public, the other salespeople should be more willing to accept the procedural change.

Many credit collections employees do not operate with computer terminals; instead, they call customers based on paper files of receivables records. If the controller changes the cash recording system so that receivables balances are reduced the moment cash is collected, it still does not improve the efficiency of the collections staff unless they have access to this information immediately—the best way to do this is to provide the entire collections staff with computer terminals. A less desirable solution is to print out paper reports for the collections staff immediately after cash is applied to receivables balances each day.

Calling Customers' Payables Departments Prior to Invoice Due Dates

This procedure is a considerable change for both the collections staff and the customers. The customers are not used to being contacted when there may be nothing wrong with the invoice and related paperwork, sometimes resulting in annoyed payables clerks at customer sites. This problem can be avoided by contacting the managers of customer accounts payable departments to explain the new system and to point out who will be calling. An introductory phone call from the collections clerk is also helpful. The company's collections staff will also protest the change, since they will rightfully claim that there is not enough time to make collections calls about overdue payments, much less to the customers whose payments are not yet due. One way around this problem is to temporarily use outside collections agencies to collect the most troublesome overdue accounts while the in-house staff switches to calling the customers with invoices that are not yet due. Another, but less useful, approach is to hire more staff to make the additional calls—the problem with this approach is that, in the long term, the number of collections staff needed will decline, so the newly hired staff must then be either let go or reassigned within the company. The better solution is temporarily using outside help.

Eliminating All Invoices

The elimination of invoices presents some tricky implementation issues. One problem is that some computerized accounting systems require invoices to be printed before they will process sales transactions, so the controller must be prepared to print invoices even if the invoices are not going to be mailed to the customers. This printing of invoices adds an unnecessary step, which slows down the overall process. Another issue is that it can be extremely difficult to match the company's record of shipments to the payments by customers, since the customers no longer match payment amounts to invoice numbers on the remittance advice. The most common way in which the customer identifies payments under a no-invoice system is to reference the purchase order number under which the parts were procured. Since the customer is using the purchase order number, it makes sense for the company also to record shipments by referencing the purchase order number. Unfortunately, very few accounting software packages allow this, so the only options are to customize the software package, manually track the shipment of goods and their related payments, or purchase entirely new software that does allow tracking of payments by purchase order number. A cheap alternative is to select an unused field in the accounting database and include the purchase order number in that field. Minor programming changes can then effect the use of that field to match payments to shipments; of course, whenever updates to packaged software are made, the programming changes will have to be redone because the update may have deleted the programmed changes.

In summary, there will be problems with implementing nearly every change advocated in this chapter. Some implementation problems are severe enough to require the replacement of an entire accounting software package, while other changes require a change of attitude by the departments involved—sometimes the personnel problems are more difficult to deal with than the technical issues.

SUMMARY

This chapter has highlighted a number of new and old technologies, reports, and measurement systems that, when used together, will increase the processing speed of the sales and receivables transaction cycle. These changes involve reducing the move and wait times associated with the transaction cycle, reducing or forestalling the number of errors in the process, eliminating processing steps, and automating portions of the process. When all these changes are in place, the accounting staff will find that sales and receivables transactions can be completed more quickly and with fewer errors.

REFERENCES

Coburn, Steve, Hugh Grove, and Cynthia Fukami. "Benchmarking with ABCM." *Management Accounting,* January 1995, pp. 56–60.

Dun & Bradstreet Business Education Services. *Collecting Past-Due Accounts.* New York, 1993.

―――. *Credit and Financial Analysis.* New York, 1993.

Hackett Group. *AICPA/THG Benchmark Study: Results Update and Analysis.* 1994.

Whittington, O. Ray, and Kurt Pany. *Principles of Auditing.* Chicago: Irwin, 1995.

Willson, James D., Janice M. Roehl-Anderson, and Steven Bragg. *Controllership.* 5th ed. New York: Wiley, 1995.

3

INVENTORY

The process of speeding up inventory-related transactions must be undertaken carefully, for inventory is frequently the largest current asset listed in a company's financial statements and usually the current asset that is most likely to be incorrectly valued. Thus, any changes in accounting for inventory will have an appreciable effect on the value of a company's largest current asset. This chapter examines the transactions most commonly used to track inventory as well as to conduct inventory counts, and then reviews several techniques that reduce both the cycle time and the processing costs associated with inventory transactions.

CURRENT SYSTEM

Inventory transactions begin when a part arrives at the receiving dock. At that time, the part is reviewed against minimum quality standards and checked to ensure that it is the correct type of part and that the quantity is correct. Then its arrival is recorded in a receiving log. The part is then moved to the warehouse; a warehouse employee signs for the part, puts it in a warehouse bin, and records the transaction. A warehouse employee later assembles (kits) several parts together in accordance with a bill of materials or sales order and moves the parts to the production area. A production employee signs for the materials on a prenumbered issuance ticket. Additional parts may be requisitioned from the warehouse small parts counter, and excess materials may be returned to the warehouse after a job has been completed; these "excess" transactions require similar paperwork. The preceding transactions may be recorded in a computer database or a manual database (e.g., on cards), or not recorded at all.

The common problem associated with this process is that a large number of transactions are required to bring a part to the production area. Each transaction increases the possibility of a record-keeping error, and a great deal of labor is needed to record the transactions without any value being added to the final product. Also,

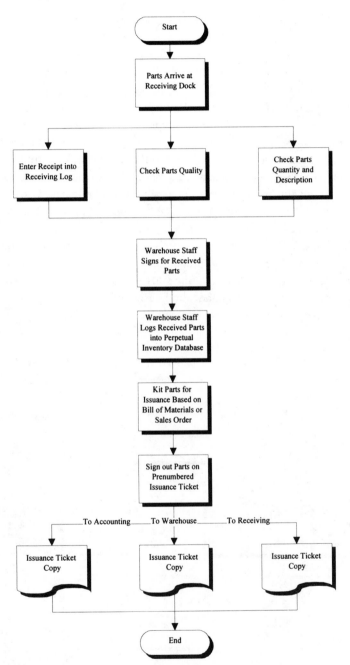

FIGURE 3.1 Sequence of inventory transactions.

recording the transactions lengthens the time needed to bring a part to the production area and requires a battalion of accounting clerks to get it there.

If the manufacturing operation is a process flow (e.g., a petroleum refinery), there may be no transactions at all. However, most traditional inventory systems follow this flow (see Figure 3.1).

Once an item reaches the production area, a cost accountant uses actual job costing to determine the total cost of the project. This involves collecting information from the warehouse about parts issued to or returned from the job, from the production staff about hours worked on the job (usually on time cards), and from several sources regarding scrapped materials. A team of accounting clerks then summarizes this information, applies an overhead rate based on one or more rate bases (e.g., the total production labor cost), and issues an interim job cost report periodically or a summary report when the job is completed. This process is shown in Figure 3.2. The traditional method of recording job costs requires considerable time on the part of the production staff (to record hours worked and materials scrapped) and the accounting staff (to record the job-related information created by the production and warehouse staffs). This reduces the productive time of the production, warehouse, and accounting staffs.

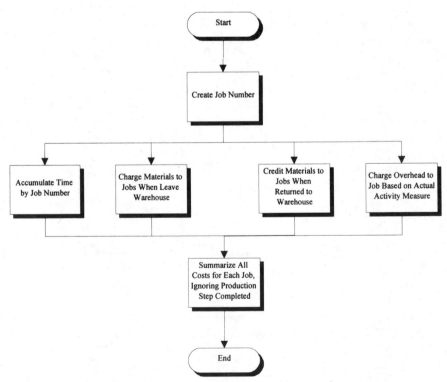

FIGURE 3.2   Actual job costing process.

Another time-consuming process is taking period-end inventory. Typically managed by the controller or an assistant controller, it begins with a definition of the roles of counting teams, early segregation of customer-owned inventory, partitioning the warehouse into counting areas, and designing and distributing inventory-taking forms. Then production must be stopped, items on the shipping and receiving docks segregated, and items counted (with a count tag left on the item counted). Next, a review team spot-checks the inventory counts, the count tags are collected, and the prenumbered tags are reviewed to verify that all tags have been collected; if not, then the warehouse is searched for the missing tags. Then the period-end counts are reviewed against perpetual inventory records (if any), and large variances are investigated. Finally, the inventory is valued, and items are reviewed for errors before a final period-end inventory report is released. A flowchart of the process is shown in Figure 3.3.

The physical inventory count is subject to a great many errors, such as miscounts, incorrect units of measure, incorrect counts by nonwarehouse employees who misidentify parts, and incorrect tag information entered by clerks into a summary database. Also, the production process is stopped while the inventory count is conducted, which may lead to lost revenue.

Another set of problems is associated with multiple inventory databases (see Figure 3.4), which must be combined to derive total inventory value. For example, a company may have a physical inventory, a separate perpetual inventory system (e.g., a card file) to track the movement of the inventory quantities in and out of the warehouse, a database of inventory costs that is not linked to inventory quantities, and a general ledger that is not linked to any of the other information. This kind of scattered information causes many problems when information changes must be recorded in the separate databases. All of the following problems result in book-to-physical adjustments:

- An item can be physically issued from the warehouse but not deleted from the physical inventory database. Alternatively, an item can be deleted from the database but still be in the warehouse.

- An item can be received in the warehouse but never added to the physical inventory database. Alternatively, an item can be added to the database but not received in the warehouse.

- The cost of an item may have changed, but the new cost is not applied against the quantity in inventory. Alternatively, an item can be removed from the warehouse, and its old cost will still be reflected in the cost database.

- All the preceding changes can occur but never be recorded in the general ledger, or they may be recorded in the general ledger even though they never occurred.

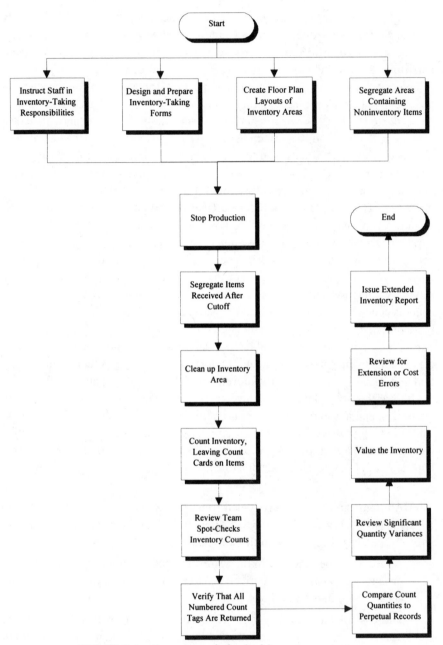

FIGURE 3.3   Sequence of physical inventory events.

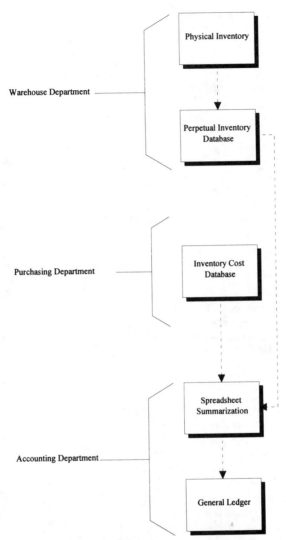

FIGURE 3.4   Multiple inventory databases.

A review of the previous flowcharts reveals a great many paperwork transfers between various employees, which add time and the risk of lost or misinterpreted information to the inventory process. An inventory value-added analysis is shown in Table 3.1. In this analysis, a value-added item is considered to be one that brings the inventory transaction closer to conclusion. The table contains estimated wait times from kitted inventory and stored inventory parts. In actual situations, this wait time can stretch to many more days, depending on inventory turnover and how rapidly the production crew is processing current jobs.

TABLE 3.1   Inventory Value-Added Analysis

Step	Activity	Time Required (Minutes)	Type of Activity
[Parts arrive at receiving dock.]			
1	Inspect parts for quality, quantity, description.	10	Non-value-added
2	Report problems to supervisor, who accepts/rejects order.	3	Non-value-added
[Shipment is accepted.]			
3	Move problem items to a parts review area.	2	Move
4	Store problem items in a parts review area.	2	Non-value-added
5	Summarize quantity of remaining items on receiving form.	5	Non-value-added
6	Move to copier.	2	Move
7	Make two copies of completed receiving form.	1	Non-value-added
8	Move to mailboxes.	1	Move
9	Put one copy of receiving form in accounting department mailbox.	1	Non-value-added
10	Move back to receiving department.	2	Move
11	Move remainder of received shipment to warehouse.	4	Move
12	Warehouse clerk inspects received shipment.	5	Non-value-added
13	Warehouse clerk signs receiving document and takes possession of inventory.	1	Non-value-added
14	Receiving clerk files signed copy of receiving document.	1	Non-value-added
15	Warehouse clerk files copy of receiving document.	1	Non-value-added
16	Warehouse clerk generates part number labels, and labels all new inventory.	10	Non-value-added
17	Warehouse clerk moves new inventory to warehouse storage area.	2	Move
18	Warehouse clerk puts inventory in warehouse bins.	12	Non-value-added
19	Warehouse clerk records location and quantity of all new inventory.	2	Non-value-added
20	Warehouse clerk moves to computer terminal.	1	Move
21	Warehouse clerk enters inventory item number, bin location, and quantity into computer.	5	Non-value-added
22	Stored items wait until needed for production.	[5 days]	Wait

TABLE 3.1 *(Continued)*

Step	Activity	Time Required (Minutes)	Type of Activity
[Production begins that requires inventory.]			
23	Warehouse clerk receives pick list from production control department.	1	Non-value-added
24	Warehouse clerk picks requested inventory items from shelf.	20	Non-value-added
25	Warehouse clerk moves to computer terminal.	1	Move
26	Warehouse clerk enters picking information into computer to take inventory items from perpetual inventory database and charge against production job.	8	Non-value-added
27	Picked items wait until needed by production staff.	[1 day]	Wait
28	Production staff compares pick list quantities to amount kitted by warehouse clerk.	5	Non-value-added
29	Production staff signs pick list.	1	Non-value-added
30	Warehouse clerk files signed pick list.	1	Non-value-added

Table 3.2 provides a summary of the value-added analysis. It shows that none of the steps bring the inventory transaction closer to conclusion; all are related to moving paperwork from person to person, inspecting the inventory, or making file copies. When the only value-added step is getting the inventory to the production area, all these steps merely delay that move. In terms of time required, the process can be entirely eliminated, while the moving, waiting, and non-value-added steps take up over eight days. In short, the actions needed to conclude the transaction are a zero proportion of the total process.

In summary, many transactions are required to move a part from the receiving dock to the warehouse, and from there to the production area. In addition, information must be collected to track job costs while an item is being produced, and

TABLE 3.2 Summary of Inventory Value-Added Analysis

Type of Activity	No. of Activities	Percentage Distribution	No. of Hours	Percentage Distribution
Value-added	0	0%	0	0%
Wait	2	6	64.00	97
Move	8	27	.25	0
Non-value-added	20	67	1.58	3
Total	30	100%	65.83	100%

yet another set of transactions is needed to conduct a physical inventory at period-end. All these transactions combine to slow down the accounting and manufacturing processes. The next section discusses a revised inventory system.

REVISED SYSTEM

Many of the transactions noted in the previous section can be eliminated by linking the cost and quantity databases as well as by removing the receiving and warehousing functions from the production process. Also, using accurate bills of materials reduces the number of inventory-related transactions. Thus, by eliminating entire transactions rather than focusing on speeding them up, the controller will find that inventory-related accounting can be completed very quickly and with less effort.

The controller can *combine the quantity and actual cost databases* involved with inventory, so that the accounting staff is not burdened with the chore of manually linking information and cross-checking it for accuracy. In addition, inventory calculation time shrinks to the time required to print out an inventory report and scan it for obvious errors (such as no cost being posted for an inventory item). A combined system of this type does not eliminate all reconciliation problems, but it does reduce them to just two items: physical inventory transactions that are not recorded in the centralized database, and transactions recorded in the centralized database that never happened in the physical inventory.

Many commercial packages link the actual cost (defined as either the invoiced cost or the purchase order cost) and the inventory quantity, but customized software may require considerable programming effort to create this link. Controls should be built into the software, so that the cost database is referenced when an inventory item is entered in order to discover items with no associated cost. Unless these items are flagged by the computer system, the costed inventory may have a number of items with no cost. An example of a centralized database is shown in Figure 3.5.

Eliminating or reducing the receiving and warehousing functions reduces the number of transactions that must be recorded and the time it takes for received parts to reach the production area. The following methods can be used to shrink the receiving and warehousing functions.

Move Fasteners to the Production Area

All low-value fittings and fasteners should be moved out of the warehouse and into the production area. This reduces the scope of cycle counting, since as much as half of an inventory's total number of parts can be fittings and fasteners. Also, the number of counts for inventory audits can be reduced, since the total number

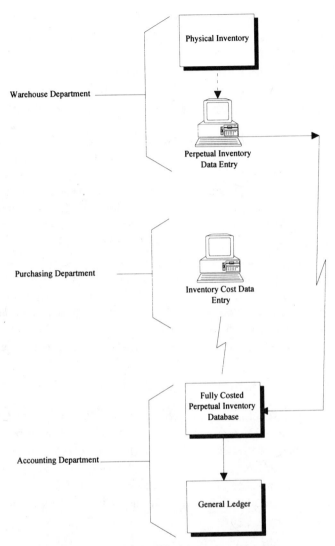

FIGURE 3.5   Centralized inventory database.

of parts in stock has been reduced. Finally, suppliers can review the bins in the production area and fill the bins themselves without anything more than a blanket purchase order; this reduces the quantity of purchasing paperwork and eliminates the receiving functions for these items.

Pay Suppliers Based on Production Records

Suppliers are usually paid based on receiving records, and this is one of the main reasons for the existence of a receiving department. If the basis of payments is moved downstream to the production department, there is less need for a receiving department. Paying based on receiving records requires precertification of suppliers for part quality and delivery times as well as very accurate bills of materials (since they are used to calculate the quantities of parts used).

Require a Purchase Order for Incoming Items

If all received items must have an associated purchase order, those incoming items without one can be segregated immediately for review. This procedure immediately isolates any customer-owned items, since they are not linked to purchase orders. By identifying and segregating customer-owned inventory immediately upon arrival, the controller can avoid considerable effort needed to identify such items when they have already been mixed into company-owned inventory.

Also, the reason for a cut-off procedure is to prevent items from being recorded in inventory without an associated cost (or vice versa). If all incoming items must have a purchase order number (which is already listed in the company's computer system with a related cost), then all inventory items must, by definition, have an associated cost that will automatically be rolled up into the general ledger. This means that the period-end cut-off procedure can be removed without any notable effect on the company's income statement or balance sheet.

Eliminate Obsolete Inventory

Every company must set aside an obsolete inventory reserve but often does not actively identify and eliminate items in inventory that are responsible for the reserve. By focusing on the number of items in inventory and setting goals to shrink it, the company can reduce inventory to only the most essential items, thereby cutting the number of cycle counts required, the number of inventory audit counts needed, and the number of inventory items to keep track of in the accounting database.

Use Just-in-Time (JIT) Manufacturing Techniques

A completely implemented JIT manufacturing system reduces or eliminates the inventory and warehousing functions by keeping only enough materials in stock for the daily production schedule; all parts for production are delivered daily by suppliers.

If all of the preceding methods are implemented, the set of transactions used to move parts from the receiving dock to the production area changes to the flow shown in Figure 3.6.

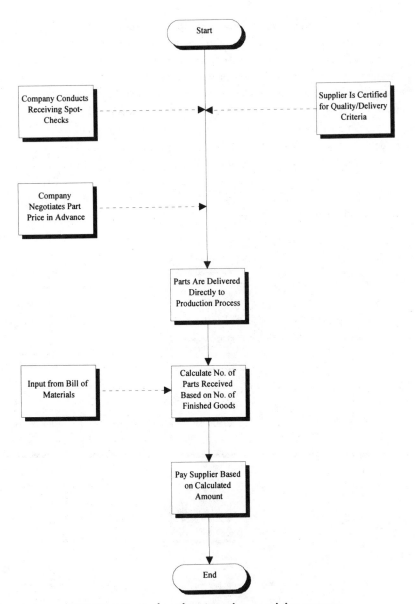

FIGURE 3.6   Reduced-transaction receiving process.

The controller may invest a large number of resources in recording the costs attached to work-in-process (WIP) production. Rather than collecting a large number of materials and labor costs from the warehouse and production areas, it is easier to *use highly accurate bills of materials to cost products* and get rid of much of the transaction reporting associated with accumulating actual job costs.

The controller can focus on keeping the bills of materials (BOMs) accurate, so that any differences between accumulated job costs and the BOMs are minimal. This removes a great deal of data entry from the accounting function and allows the production crew to manufacture products instead of reporting on them. Best of all, the cost accountant can review the bills of materials before a product is even produced, to determine where costs can be saved. Actual job costing does not allow the cost accountant to do this, since cost reports focus on costs that have already been incurred, which places the cost accountant in a reactive mode instead of a proactive mode. Job costing using bills of materials is shown in Figure 3.7.

If *perpetual inventory records are maintained with high accuracy,* no physical inventory should be conducted. An increasing number of companies are turning to this method for the following reasons.

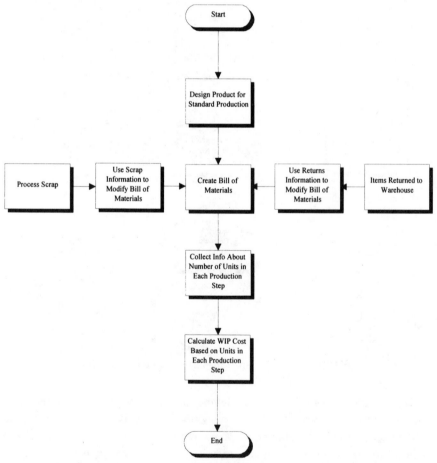

FIGURE 3.7   Job costing using bills of materials.

Avoid Wasted Time

Staff time is not efficiently utilized during the physical inventory process, because the employees could be involved in other activities. Also, the production facility is shut down, preventing revenue-generating products from being manufactured.

Improve Product Delivery Performance and Reduce Freight Costs

Accurate inventory records allow the company to promise shipments to customers with greater confidence, because products can be built without delays due to missing parts. Also, rush charges for missing parts are avoided.

Achieve Higher Accuracy with Ongoing Counts

Inventory counts should be done by the experts—the warehouse staff—and should be done at their leisure, which ensures higher count accuracy. If a complete plantwide physical inventory is performed, accuracy drops because of counts by less experienced nonwarehouse staff and the short time period available to complete the count.

Avoid Year-End Surprises

Many companies have been unpleasantly surprised by unexpected changes in inventory levels at year-end. These surprises can be avoided by constantly monitoring inventory levels with a perpetual inventory system.

Use Data to Reduce Inventory and Cut Costs

The transaction history that is a by-product of a perpetual inventory system allows the materials manager to make informed decisions regarding deletions of parts from stock. This is of value to the controller, since cash requirements for additional inventory are reduced and can be enhanced as inventory is sold back to suppliers. As inventory is reduced, the staff needed to track it and the insurance needed to insure it can both be reduced, thereby improving the company's cash flow a second time.

The controller should realize that a perpetual inventory system requires constant review to keep it accurate. Thus, even after the system has been installed, cycle counting and inventory audits must become a daily or weekly part of the warehouse and auditing departments' work load.

In summary, the controller can greatly reduce the number of inventory-related transactions by maintaining highly accurate bills of materials to cost products and pay suppliers, and using purchase orders to identify customer-owned inventory, eliminate the period-end cut-off procedure, and automatically cost inventory. A variety of techniques can be used to eliminate or reduce the inventory, thereby reducing the number of transactions related to the warehousing function. Though it is not always possible to completely eliminate the receiving and warehousing functions, a simple reduction in the number of transactions associated with them will allow the controller to focus more attention on the smaller number of transactions that are being used as well as to divert accounting resources into other, more critical activities.

CONTROL ISSUES

Many control issues are related to inventory. For example, if inventory were not physically controlled with an inventory cage, employee pilfering might occur, resulting in write-downs. Similarly, removing controls over the recording of inventory receipts increases the risk of inventory being pilfered before it even reaches the warehouse storage area. This section contains examples of what would happen if various inventory controls were removed in order to increase the speed of inventory transactions, and it suggests possible solutions to the resulting control problems. These examples pertain to the changes suggested previously for a revised inventory system.

Do Not Conduct a Physical Inventory

This control is used to verify that the company's book inventory balance matches the physical inventory balance. If it were removed, the risk of the company's book balance varying appreciably from the physical balance would be high.

Create a Perpetual Inventory System.   A perpetual system allows the company to keep inventory balances up-to-date and matched to the physical inventory amounts at all times. Since the inventory is being tracked constantly and reviewed with cycle counting and periodic inventory audits, there is no need for a physical inventory count. This method eliminates one of the more common forms of inventory-related fraud, which is the mislabeling of parts, use of wrong units of measure, and counting of empty boxes as legitimate inventory—continual audits discover these practices quickly and tend to discourage them.

Use a JIT System to Reduce Inventory.   If there is no inventory, then one does not have to count it. A just-in-time (JIT) manufacturing system tends to reduce the amount of inventory on hand, which may eventually result in an inventory so small that it is numerically insignificant and does not require periodic inventory counts.

Stop Valuing Inventory with Standard Costs

Standard costs provide a valuation baseline for all parts recorded in the inventory. If the control is removed, the potential exists for some items to have a zero cost. On the other hand, standard cost can be abused since it is subject to change by any authorized user. For example, a controller who wants to improve reported company profits could incrementally alter a large number of items so that individual costs look reasonable but summarize into a large increase in the inventory balance, resulting in a smaller cost of goods sold and a larger profit.

Use Actual Costs. By linking the purchase order cost database to the inventory quantity database, actual costs are already recorded for each item before it is received (since the purchase order must be completed before the item can be ordered). Thus, the cost is not subject to arbitrary change by a user, and all parts will be costed.

Stop Including Floor Stock in Inventory

Floor stock turns over constantly and usually involves low-cost items. It is very difficult to include in a perpetual inventory, since it is hard to keep an accurate count of the items. If the control is removed, pilfering of stock becomes easy—after all, no one is reviewing the amount of inventory that is left out in the open, within easy reach of all production staff.

Conduct a Cost/Benefit Analysis. Some floor stock items are so expensive that they must be kept in a locked area. However, most floor stock items (e.g., nuts, bolts, and washers) are so inexpensive that it is easier for the company to order them in large quantities and leave them out in the open, giving everyone access to the materials. A simple analysis will conclude that it is cheaper to leave low-cost items in the open than to track individual inventory transactions every time a floor stock item is used.

Stop Reviewing Inventory for Customer Ownership

Accountants are often called upon to review the inventory and ensure that no items included in the inventory compilation actually belong to someone else. If the review never occurs, employees could mix this inventory into company-owned inventory, thereby artificially increasing the inventory balance and lowering the cost of goods sold.

Create a Segregated Area for the Goods.   Create a locked area for customer-owned materials, into which goods are taken directly from the receiving area. This prevents customer-owned inventory from being mixed into company-owned inventory.

Shift Control to the Receiving Function.   As usual, one of the best ways to eliminate a control is to create a new control somewhere else that offsets the loss of the first one. It is nearly impossible to determine ownership once customer-owned inventory has been placed in the warehouse. A more appropriate (and less time-consuming) control point is to enforce the receipt of goods with purchase orders. If an incoming item has no purchase order (which should include virtually all materials owned by someone else—after all, the point of a purchase order is to establish the company's intent to buy the product), it should be set aside for review by the internal audit or accounting staffs, who can quickly determine whether it is customer-owned. The item can then be marked as customer-owned and moved directly to a segregated area for customer-owned materials. This eliminates one of the more common types of inventory fraud, which is recording customer-owned inventory as though it were the property of the company.

Eliminate the Warehouse

A warehouse has two uses: it tracks excess materials, and it provides information to the controller about actual materials costing in the form of issuance tickets for all items withdrawn from the warehouse. If the warehouse were to be eliminated, there would be no custodial control over inventory, and the controller would not have any materials costing information.

Eliminate Excess Materials.   If the company can eliminate unused items, return excess items to suppliers, and move fittings and fasteners out to the shop floor for easier access by the production staff, there is no need for a centralized storage facility. Very expensive items that cannot be ordered on a just-in-time basis should still be stored in a locked facility, but the company's overall storage needs should still shrink considerably.

Use Bills of Materials.   If the company has accurate bills of materials, it already knows the cost of materials used in its products and has no need for warehouse records regarding materials issuances to specific jobs. However, the bills must be highly accurate for this kind of costing system to work effectively.

Stop Collecting Actual Work-in-Process Costs

Actual work-in-process costs are used to summarize a job or production run after it has been completed, so that management can review costs and decide if any action is needed to improve the situation if the production will occur again. This cost summary would not be possible if the data collection were to stop.

Use Bills of Materials. Tasks should only be completed once. Since nearly all companies are (or should be) completing a bill of materials for each product manufactured, the information in a job costing summary is already available. An objection might be that this bill of materials information is a theoretical standard, which is not comparable to actual production costs. However, bills of materials are constantly updated with information about actual scrap rates, missing parts, and excess parts returned to the warehouse. These updates must occur, since the bill of materials must be accurate enough to use for purchasing, kitting, and (if used) a materials requirements planning (MRP) system. As for labor rates, the bill of materials is also used to determine production capacities (especially if a manufacturing resources planning system is used), so the bill of materials should also reflect accurate labor quantities. In short, an accurate bill of materials should yield costs similar to those of an actual production cost collection system.

Stop Receiving

The receiving function is critical for ensuring that the correct quantity and type of items are received, and that their quality meets company standards. This department also logs receipts into the computer system; since the logged-in receipts may be automatically matched with purchase order prices and scheduled for payment, the receiving function becomes especially critical. Thus, if the function were to be eliminated, the controller would have some difficulty ensuring that items were received and that a proper period-end cut-off had occurred.

Certify Suppliers. The engineering and purchasing departments can certify suppliers for quality, on-time shipping reliability and the correct number of parts delivered. Once suppliers are certified, one need only check received items occasionally to ensure their quality.

Pay Suppliers Based on Production Records. In accordance with just-in-time manufacturing concepts, the company should have suppliers deliver parts directly to the production process. Then, once production is completed, the company can determine the number of supplier parts used by multiplying the number of finished goods by the part quantities listed on the bill of materials. A key to this process is not having any excess parts delivered, since this back-flushing technique ignores any extra parts that were not used to produce finished goods.

Stop Reviewing the Period-End Cut-off

In a traditional accounting system, adherence to a strictly defined cut-off date at the end of each accounting period is mandatory. The accounting staff spends a considerable amount of time each period ensuring that this cut-off is being observed. The reason for this cut-off is that if an item is received into inventory at month-end but its related invoice is not recorded, inventory would increase without a related cost, which would reduce the cost of goods sold and therefore increase profits. Alternatively, an item might not be added to inventory though its related supplier invoice might be logged in as an expense, thereby increasing the cost of goods sold and decreasing profits. However, the period-end cut-off does not need to be reviewed by the accounting staff in a revised inventory system.

Only Receive Items If a Purchase Order Already Exists.   The company's computer system must only allow items to be received if a purchase order (with a unit cost) already exists in the system. Then, as long as the controller strictly enforces the entry of receiving information into the computer system, it will be impossible for items to be received into inventory without a matching cost automatically being recorded at the same time by the computer software. Even if the period-end cut-off is being followed in a sloppy manner, there will be no effect on profits. This eliminates fraud related to recording the inventory item but not the related payable in order to increase the inventory valuation. However, the controller must still watch for items that may be double-counted if they are in transit between two company locations (the risk is best avoided if there is a single inventory database for both locations).

All inventory controls should not be eliminated. Some controls are still necessary, since they preserve the segregation of duties, affect the completeness of accounting paperwork, ensure that management reviews key transactions, or ensure the physical segregation of inventory. In particular, the following controls should be maintained to avoid the potential increase in risk associated with their elimination.

Establish Physical Control over Inventory

Unless a company decides to eliminate inventory with a just-in-time manufacturing system, the inventory must be controlled in a fenced-in, locked area. In addition, an employee must be given total responsibility for the accuracy of the inventory (as well as for any inventory shortages in the warehouse), including being in charge of recording all movements of stock into and out of the inventory area. Otherwise, too many people will be able to enter inventory transactions into the company's inventory records, usually resulting in inaccurate information. Some warehouse operations even require a signature from the warehouse staff person who handles each transaction, though a more automated approach is to record the user identification of a computer operator each time a transaction is entered. Finally, only warehouse personnel should be allowed into the warehouse, with visitors requiring an escort—this keeps inventory pilferage to a minimum.

Track Obsolete Inventory

The controller must be aware of the number and cost of items in the inventory that may be obsolete. A good way to derive this information is to track the last time an inventory transaction occurred for an inventory item (a data element commonly found in a perpetual inventory system). A report dealing with this issue is shown in Figure 3.11.

Audit the Inventory

The controller should institute reviews of the inventory information by someone who does not report to the inventory manager. The ideal person would be an internal auditor; in the absence of such a person, the cost accountant or another inventory-knowledgeable person should conduct the reviews. If a perpetual inventory is available, the review should be conducted weekly. The reviewer should take a random sample from the inventory report (sorted by location and including extended costs) and review the following:

- Items physically in stock are listed on the report.
- Items listed on the report are physically in stock.
- Units of measure are correct.
- Large-dollar items are correct (sample 100% of these items).

The inventory report from the previous week should also be retained, so that the following can be reviewed:

- The total number of part numbers in stock does not vary much from the previous week; if it does, such variances can be identified and explained.
- The total dollar value of inventory does not vary much from the previous week; if it does, such variances can be identified and explained.

Segregate the Purchasing and Inventory-Keeping Functions

It is important to continue to segregate these functions, for otherwise an employee could purchase an item and then divert the incoming materials for personal use.

Account for the Sequence of Purchasing Documents

For companies who rely on the purchase requisition or purchase order form as the primary control over the purchase of materials or services, the theft of either of these forms is a serious breach of control. These forms could be used by unauthorized staff to purchase materials in the name of the company, with shipment to some other location. The company would only discover the theft after being billed by the supplier for the goods.

Authorize Purchasing Documents

A case can be made for reducing the number of approvals for materials (especially for low-cost items). However, some approval must be maintained for the more expensive items to prevent employees from purchasing items that may impede the company's cash flow or even require an excessive amount of corporate financing to purchase.

Review Parts Costs by Supplier

Under a just-in-time manufacturing system, inventory is minimized and consequently the risk of fraud is greatly reduced. However, to avoid collusion between buyers and suppliers (who no longer have to bid for work), the controller should periodically review supplier costs versus a sample of costs from other suppliers. The controller should also review prices charged by suppliers versus contracted prices.

Use a Receiving Log

A key control item is the receiving log. Every received item must be listed in it. If an item is not noted in the receiving log, the controller has no way of knowing if an item has been received and therefore cannot pay a supplier for the item without first receiving a proof-of-delivery document from the shipper (a time-consuming process that does not engender good relations with a company's suppliers). A computerized receiving department would enter the receiving information into a computer database rather than manually into a paper log.

Conduct Closed Job Reviews

One of the more useful control techniques is the after-the-fact review of closed jobs or production runs. Aided by a summary of costs, the management team should review each project and summarize the items needing improvement; this information is useful if the job is to be repeated. This technique is useful for keeping management mistakes from occurring a second time. The controller has a central role in this process, since the accounting department must provide all relevant project costs to the review team.

Use Accurate Bills of Materials

One of the most critical records is the bill of materials, which lists the amounts and types of parts needed to build a product. When accurately maintained, it allows a company to predict the correct types and quantities of parts to purchase for an upcoming production run (e.g., MRP). More important, it allows a company to remove items from its perpetual inventory records once a product has been completed (called back-flushing). If the bills of materials are incomplete, MRP and back-flushing cannot be used.

To control the bill of materials information, cross-check additional items that are requested from stock during a job's production or returned to stock following a job's production. These transactions indicate that a product's bill of materials is incorrect. Also, distribute forms to the production and kitting staffs, who can report on missing or duplicate parts. Finally, a review committee can systematically review all bills of materials for inaccurate quantities, part numbers, and units of measure. Since the controller bases year-end inventory projections on product costs, which in turn are based on bills of materials, correcting the bills can eliminate the annoyance of large year-end inventory variances.

Use Accurate Scrap Information

Of particular concern when using a bill of materials to back-flush an inventory is the assumed scrap rate built into the bill of materials—if it varies a lot from the actual scrap rate, actual inventory may vary substantially from the book balance. Also, if a company's bill of materials software only allows one scrap rate for everything listed in a bill, then back-flushing may result in incorrect inventory balances, because scrap rates usually vary by component within a bill of materials.

In summary, controls surrounding the storage and movement of inventory can greatly increase the time required to complete inventory-related transactions. However, by adopting new systems such as a perpetual inventory, just-in-time techniques that bypass the receiving and warehouse functions, and replacing actual cost collection with accurate bills of materials, the controller can reduce the time required to process inventory-related transactions.

QUALITY ISSUES

Inventory-related accounting transactions can be slowed considerably by the introduction of mistakes into the process. Thus, high-quality inventory processing is defined as processing with the smallest number of mistakes. This section notes the most common inventory-related mistakes and suggests how to prevent them. Some errors listed here are not necessarily caused by accounting personnel, but they may have to be researched and fixed by the accounting staff, so they are included in this section.

Incorrect Recording of Units of Measure

Different departments prefer different units of measure for parts. For example, the warehouse may prefer to count sheet metal in sheets, but the purchasing department may order it in pounds, and the engineering department may list it in square feet on product drawings. Thus, if an unauthorized person accesses an inventory record and changes the unit of measure, the quantity on hand can change drastically. For example, a 400-pound sheet of metal that costs $825 can become 400 sheets of metal that cost $330,000.

Incorrect Recording of Quantities

A simple input error can result in an incorrect quantity in the inventory database. This error can be made by either the warehouse or receiving staffs. The worst result of this is not having enough parts on hand when they are needed, thereby delaying production while more are procured.

Incorrect Recording of Locations

When an inventory location code is incorrectly input into the inventory database, the inventory is "lost" even though it is still in the warehouse. If the warehouse is very large, it may be more cost-effective to reorder the lost part rather than undertake an extensive search for it. Cycle counting or a random audit check will eventually find the part, but this may take a long time.

Incorrect Recording of Purchase Order Costs

A part cost may be incorrectly input into the purchase order database, which may then be used to cost the inventory. This is not an error that usually stays uncorrected, because the supplier will call to complain if the cost is too low. However, the supplier may not complain if the cost is too high and he is therefore paid too much money.

Incorrect Recording of Changes to
Bills of Materials (BOMs)

There are two errors related to BOMs: changes are made to the BOM database that are not reflected in actual usage, and changes are made to the product without updating the BOM database. The first error results in extra parts being purchased based on the incorrect BOM data, and the other results in items not being purchased, since the information is not listed in the BOM. The first error results in extra parts arriving at the warehouse, and the other error results in production delays while parts are procured on a rush basis.

The quality of inventory-related transactions can be improved in various ways. Supervisory reviews or duplicate data entry are not very useful ways to improve quality, because these methods simply review transactions for errors that have already been made. The best method is to keep the errors from occurring in the first place by reducing, eliminating, or simplifying manual, error-prone labor. The following procedures will improve the quality of inventory-related transactions.

Limit Access to Unit of Measure Information

Only an authorized engineer and a direct supervisor should have access to the computer database records for parts, thereby preventing unauthorized changes of units of measure.

Provide Multiple Units of Measure for Parts

Some manufacturing software provides a table for listing multiple units of measure for each part. For example, sheet metal can be listed in sheets, pounds, or square feet. No matter which unit of measure is used, the transaction will still be accurate.

Description	Unit of Measure	Quantity
Sheet, Stainless, 316L, 7 GA	lbs	378
	sheet	1
	sq ft	48

Eliminate the Entry of Any Inventory Quantities

The best way to avoid transaction errors is not to have any transactions. This can be achieved by having incoming shipments sent directly to the production area for immediate inclusion in products, thereby avoiding the receiving and warehouse functions. Suppliers are paid based on the number of units produced by the company, which implies that only the supporting bill of materials must be accurate. All other receipt, issue, and cycle count adjustment entries are avoided.

Use Bar Codes for Data Entry

Arrangements can be made with suppliers to mark quantities on parts shipments with bar codes, so that the receiving and warehouse clerks can scan the information directly into the computer database without risk of causing a transaction error. Also, warehouse location codes can be set up as bar codes, so that key-punch errors can be avoided in this area as well.

Use Computer Limit Checks

The computer can be programmed to reject transactions if information falls outside of preset ranges. For example, purchase order costs can be rejected if they vary notably from previous costs already in the database. Quantities can be limited based on reasonableness (e.g., if a quantity being entered would overload the entire warehouse and spill onto the street, then it should be rejected). Issue transactions can be rejected if they would result in negative inventory quantities. Receipt transactions can be rejected if they sum to an amount greater than the quantity listed on the referenced purchase order. Finally, location codes can be rejected if they are for locations that do not exist.

Cost Accountant Review of Planned Costs

The cost accountant should be part of the team that designs products, with input into the process regarding parts costs that sum to the total product cost. The cost accountant can then review costs on purchase orders, thereby checking for cost input errors. After products have been designed, a key role of the cost accountant is to continue to review purchase orders to ensure that suppliers are charging the costs that were planned when the product was designed. This review ensures that purchase order costs are not input incorrectly.

Limit Access to Bill of Materials Information

Only an authorized engineer and a direct supervisor should have access to the computer database records for bills of materials, thereby preventing unauthorized changes from occurring.

Create Overlapping Bill of Materials Change Sources

Changes to bills of materials can be reported to the bill of materials engineer by the engineer in charge of the product, by the warehouse staff that receives excess materials back into the warehouse, by the warehouse staff that issues additional parts to the production department to complete a product, and by the production staff during their assembly of a product. By creating these multiple sources of information, the bill of materials information can be kept up-to-date.

In summary, inventory-related transaction errors can be avoided by limiting access to key database records and creating multiple review sources for key information that feed changes back to authorized personnel. Also, information can be entered into the computer via bar codes, and the software can contain automatic field limit checks and multiple units of measure for each part. Finally, transactions can be avoided entirely by using just-in-time manufacturing techniques and using bills of materials to record receiving information.

COST/BENEFIT ANALYSIS

This section demonstrates how to conduct cost/benefit analyses for implementing a perpetual inventory system, removing floor stock from the warehouse, eliminating the review of customer-owned inventory, eliminating the warehouse, stopping collection of actual work-in-process costs, eliminating the receiving function, and eliminating the period-end cut-off procedure. In these examples, expected revenues and costs are as realistic as possible. All costs or benefits are treated as though they occur in just one year; this avoids the complication of multiyear cost/benefit examples.

Implement a Perpetual Inventory System

Bill Sweet, president of the Slow Times Syrup Company, wants to convert to a perpetual inventory system. He asks the controller, Mr. Honeycut, to analyze the

costs and benefits related to the conversion. Mr. Honeycut finds that the company has had an average unexplained inventory write-down of $53,000 in each of
the last five years after the year-end physical inventory was conducted. An extensive review of the bills of materials indicates that they are accurate and have
been accurate in the past. Also, the purchasing department calculated that it
spends an extra $12,000 per year on rush freight charges to bring in materials
that were supposedly in the warehouse but could not be located. The warehouse
must be enclosed for the perpetual inventory; it will cost $8,000 to fence in the
warehouse area, plus $4,500 to install a computer and link it to the accounting
software on the company network. In addition, two warehouse clerks must be
hired at salaries of $18,000 each, plus a 10% overtime premium for the clerk
working the second shift. Finally, two hourly employees must be assigned to the
warehouse for three months to help with arranging the inventory, tagging it, and
logging it into the computer; each of these employees is paid $14 per hour. Is it a
good idea to install a perpetual inventory system?

Solution.   The cost of installing a perpetual system versus the cost of *not* installing one must be determined.

Cost of Not Installing a Perpetual Inventory

Cost of unexplained inventory write-down	$53,000
Cost of rush freight charges	12,000
Total cost of nonperpetual system	$65,000

Cost of Installing a Perpetual Inventory

Cost of fencing in warehouse	$ 8,000
Cost to install a computer linked to network	4,500
Cost/yr of first-shift warehouse clerk	18,000
Cost/yr of second-shift warehouse clerk	19,800
	$50,300
Cost/hour of employees for system setup	$ 14
No. of employees for setup	× 2
Hours in three-month setup period	× 520 hrs
Total cost of employees for setup	$14,560
Total installation cost	$64,860

In short, the cost of *not* having a perpetual inventory is $65,000, whereas the cost
of installing one is $64,860, which indicates that having a perpetual inventory is
a worthwhile project. Since the cost of installing the system is only a one-time
occurrence, the benefit in future years will be greater than in the first year.

Remove Floor Stock from Warehouse

Mrs. Toadstool, the warehouse manager, has been ordered to cut her budget for the next fiscal year by 20%, which is a $10,000 cut. To do so, she has to reduce the hours of her cycle counting employee. It looks as though inventory accuracy will plummet if this happens. As the astute controller, you print out a list of inventory items, and identify 500 parts out of the total inventory of 3,000 parts that are fittings or fasteners and that cost 50 cents or less per unit. The total cost of these items is $14,000 out of a total inventory valuation of $662,000. When these items arrive, the warehouse staff uses a counting scale to bag them into clusters of 100 items per storage bag; this allows the cycle counter to count the parts more easily. The bagging process takes 5 minutes per receipt, and roughly 150 items are received per week. Parts are reordered based on a reordering report that is automatically printed every day and forwarded to the purchasing department. The cycle counter can count an average of 40 items per hour and counts the entire inventory once every two weeks. You discover that a storage bin for each floor stock item (to be placed in the middle of the shop floor) can be purchased for $8 per bin plus $5,000 for the rack (to be depreciated over ten years) and that a warehouse staff person can review the rack daily and take 2 hours to write down items that require reorders. A typical warehouse staff person is paid $12 per hour. If the bin is placed on the shop floor, pilferage is expected to be 10% per year. Is it worthwhile to move the floor stock out of the warehouse and into the production facility?

Solution.   The cost of counting floor stock in the warehouse versus the cost of maintaining it on the shop floor must be determined.

Cost of Keeping Inventory in Warehouse

Time/week bagging floor stock	12.50 hrs
Weeks/yr	× 52
Time/yr bagging floor stock	650 hrs
Labor cost/hour	× $12
Total inventory maintenance labor cost	$7,800
No. floor stock items	500
Items counted/hour	÷ 40
Time to count floor stock every two weeks	12.5 hrs
No. of times/yr inventory counted	× 26
Time/yr to count floor stock	325 hrs
Labor cost/hour	× $12
Total cycle count labor cost	$3,900

Cost of Moving Inventory to Shop Floor

Cost of floor stock	$14,000
Pilferage percentage	× 10%
Pilferage cost/yr	$ 1,400
Time/week to review stock for reorder	2 hrs
Weeks/yr	× 52
Review hours/yr	104 hrs
Cost/hour	× $12
Cost/yr to review stock for reorder	$ 1,248
No. of bins required	500
Cost/bin	× $8
Total bin cost	$ 4,000
Cost of rack for bins	5,000
Total cost of rack and bins	$ 9,000
No. of years of depreciation	÷ 10
Depreciation cost/yr	$ 900

In short, the cost of keeping floor stock in inventory is $11,700 per year, whereas the cost of keeping it on the shop floor is $3,548, a savings of $8,152 per year. This savings will allow cycle counting to continue for all other items remaining in the warehouse while reducing the warehouse budget by more than 20%.

Eliminate Review of Customer-Owned Inventory

You are the controller of CMP (Consigned Material Products) Inc. Your accounting staff is overwhelmed during physical inventory taking because they spend days figuring out which inventory in the warehouse belongs to other companies. You decide to install a control point at the receiving location to correct this problem: all incoming items without accompanying purchase order numbers will be placed in a holding area for immediate review by the accounting staff, with customer-owned items placed in an enclosed section of the warehouse. Currently, the job of reviewing the ownership of inventory every month requires three days of work by the assistant controller, who earns an annual salary of $42,000. The cost to enclose a portion of the warehouse will be $2,400. A computer terminal must be installed in the receiving area so that purchase order information will be available on line; this will cost $4,500, which includes network connections. The assistant controller will need 15 minutes each day to review possible consignment inventory at the receiving control point. Is this a cost-effective control?

Solution.   The cost of segregating customer-owned inventory is compared to the cost of identifying it without additional controls.

Cost to Segregate Customer-Owned Inventory

Cost to enclose a warehouse area	$2,400
Cost to install a computer terminal	4,500
	$6,900
Cost/hour of asst. controller	$20.19
Daily inventory review time	× .25
Daily cost of inventory review	$ 5.05
No. of business days/yr	× 260
Cost/yr of inventory review	$1,313
Total cost to segregate customer-owned inventory	$8,213

Cost to Track Customer-Owned Inventory with
No Additional Controls

Daily cost of asst. controller	$161.52
No. days/yr to review inventory	× 36
Cost/yr of inventory review	$ 5,815

The cost of adding a control point at the receiving stage is $8,213 in the first year, versus a cost of $5,815 to avoid the control point. However, once the fixed costs have been spent in the first year, the cost of having the control point drops to only $1,313 (the labor cost) in later years, versus a cost of $5,815 if the control point is not installed. Therefore, if multiple years of payback are considered, the receiving control point should be installed.

Eliminate the Warehousing Function

The president of Custom Welded Products, a manufacturer of customized glove-boxes for the nuclear industry, has reviewed the economic value added of each department and concluded that the warehouse adds no value to the finished product; therefore, it should be eliminated. As the controller, you explore the costs and benefits of this action. Fittings and fasteners can be moved to the shop floor, but 10% pilferage of the $10,000 of these items is expected. If no extra stocks are kept on hand, it is estimated that $6,500 will be needed each year to order and ship in parts from suppliers on a rush basis, even if reasonably accurate bills of materials are maintained. Also, an additional staff person must be hired in the engineering department to prepare more accurate bills of materials, so that parts are ordered in exactly the right quantities. About 25% of the materials used in the products are 20% cheaper when ordered in bulk; if the warehouse is eliminated, they must be ordered in just the right quantities to meet production requirements. The company's annual materials cost is $1,500,000. The two warehouse staff people are paid $32,000 and $18,500. Also, $5,500 of annual depreciation will be eliminated if the warehouse racks are sold off. The resale value of the racks is $18,000. The raw materials inventory valuation is $1,250,000, and the company invests its short-term funds at an interest rate of 4%. Should Custom Welded Products eliminate its warehouse?

Solution.   The benefit gained from eliminating the warehouse must be balanced against the costs associated with ordering smaller quantities, rush freight shipments, pilferage, and the extra labor to maintain accurate bills of materials.

Benefit of Eliminating Warehouse

Eliminate warehouse salaries	$ 50,500
Sell racks	18,000
Reduction in depreciation	5,500
	$ 74,000
Cost of inventory	$1,250,000
Cost of floor stock retained	− 10,000
Net savings on inventory	$1,240,000
Interest rate on investment	× 4%
Net earnings from add. working capital	$ 49,600

Cost of Eliminating Warehouse

Cost of items moved to shop floor	$ 10,000
Expected pilferage	× 10%
Total expected pilferage cost	$ 1,000
Cost of rush shipments	$ 6,500
Cost of bills of materials employee	$ 32,000
Cost/yr for materials	$1,500,000
Materials cheaper in bulk	× 25%
Cost of bulk materials	$ 375,000
Small-order-quantity surcharge	× 20%
Cost of small-order-quantity surcharge	$ 75,000

The total cost of eliminating the warehouse is $114,500, and the total benefit is $123,600. Thus, it is cost-beneficial to eliminate the warehouse.

Stop Collecting Actual Work-in-Process Costs

You, the controller of Steady State Systems, have just met with the controller of Amalgamated Products Unlimited and discovered that her accounting department is two-thirds the size of yours, even though both companies are similar in size and function. Upon further inquiry, you find that the other controller does not report on actual work-in-process costs, relying instead on bills of materials generated by the engineering department and updated with selected information gathered from the shop floor. You are determined to try this as well but must convince the top managers of your company that they will still receive high-quality cost information while you cut personnel costs. After considerable investigation, you find that bills of materials currently show costs that vary from actual costs by an average of 15%. In order to report information that will not lead to poor decision making, you must reduce this to 5%. To do so, an engineer must be hired to continually update bills of materials information; this person's salary will be $45,000. Also, the warehouse staff must report on items returned to the warehouse from the shop floor as well as on extra parts issued to the shop floor (which reveals excesses or shortages on the bills of materials). This will require a half-hour of warehouse time every day; the average warehouse worker earns $18,000 per year. In addition, the results of the existing scrap reporting system must be channeled to the engineer, who will incorporate this information into the bills of materials; there is no cost associated with this step. Finally, the bills of materials engineer must meet with the production supervisors every month to review the labor rates shown on the bills of materials and adjust them as necessary; the cost of supervisory time for this process is $8,000 per year. If the existing work-in-process cost accounting system is eliminated in favor of costs based on bills of materials, then you will be able to eliminate a cost accountant and an accounting clerk from your payroll. The cost accountant earns $35,000 per year, and the clerk earns $20,000 per year. Based on this costing information, should you switch to bills of materials costing?

Solution. The costs and benefits of the switch must be determined.

Cost of Bill of Materials Costing

Cost/yr of bills of materials engineer	$45,000
Cost of review time with supervisors	8,000
	$53,000
Average warehouse pay/hour	$ 8.65
Time/day to update BOMs	× .5 hr
Cost/day to update BOMs	$ 4.33
No. of business days/yr	× 260
Cost/yr to update BOMs	$ 1,126
Total cost for BOM costing	$54,126

Cost of Actual Work-in-Process Costing

Cost/yr of cost accountant	$35,000
Cost/yr of accounting clerk	20,000
Total pay of WIP reporting staff	$55,000

In short, the cost of starting up a bills of materials costing system is $54,126 versus a cost of $55,000 if the current actual work-in-process reporting system is maintained. Based on this information, the bills of materials costing system should be used.

Eliminate the Receiving Function

The CFO of Fonicka Cameras, Inc., a manufacturer of high-quality cameras, wants the controller to cut costs in the receiving area. The controller finds that only one of the two receiving employees will be needed if the company certifies its suppliers and has them deliver products directly to the camera production line. The company will save $18,000 by eliminating the receiving position, plus an additional $9,000 by reducing the other position to half-time, but it must still maintain a receiving computer workstation to handle deliveries by smaller suppliers who are not certified. Also, two warehouse positions, each costing $16,500 per year, will be eliminated, since no items will be processed through the warehouse. The cost to monitor and certify suppliers is $30,000 in the first year, which pays for the part-time labor of an engineer and a purchasing agent, plus their travel costs to visit suppliers. This certification cost is expected to go down to $15,000 after the first year. In addition, the cost of purchased parts is expected to be unchanged when more frequent deliveries are enforced. The cost would have been higher, but all parts will now be single-sourced, so suppliers will absorb the delivery cost in exchange for higher purchased volumes. Finally, payments to suppliers will be made based on completed production volumes; the number of parts used in each completed item is based on a bill of materials. The number of parts in each bill must be carefully reviewed to ensure proper payment, and this requires the half-time labor of a $35,000 junior engineer. Is it worthwhile to eliminate the receiving function?

Solution.   The costs and benefits associated with eliminating the receiving function must be determined.

Cost of Eliminating the Receiving Function

First-year cost to certify suppliers	$30,000
Cost to review bills of materials	17,500
Total cost of eliminating receiving function	$47,500

Cost of Retaining the Receiving Function

Cost of retained warehouse positions	$33,000
Cost of retained receiving positions	27,000
Total cost of retaining receiving function	$60,000

The receiving function has several jobs in the warehouse tied to it, so they are also eliminated if the receiving jobs are no longer needed. Thus, the total cost associated with keeping the receiving function is much higher than the cost of just the receiving payroll. This means that the cost of retaining the receiving function is $60,000 versus a cost of $47,500 to convert to a system with a reduced receiving function. Therefore, Fonicka Cameras should reduce the role of its receiving function.

Eliminate the Period-End Cut-Off Procedure

As the controller of the J. P. Murphy Company, you find that it takes a half day for the assistant controller (whose salary is $45,000) to cross-check inventory received during the period end against invoices received from suppliers, thereby ensuring that items received have a matching cost. If the receiving station had a computer terminal linked to the accounting software, then all receipts could be entered on the spot and automatically matched to purchase orders that are already in the computer database, having been previously entered by the purchasing staff. The cost to install this terminal and link it to the company's computer network is $4,500. The largest cut-off error the assistant controller ever discovered was for $78,000, when inventory was recorded on the company's books but no accompanying purchase cost was recorded. The company's total current assets average $2,000,000, and its current ratio is 2:1. Is it worthwhile to eliminate the period-end cut-off procedure?

Solution. The costs and benefits of eliminating the period-end cut-off procedure must be determined.

Cost of Eliminating the Cut-Off Procedure

Cost to install a computer terminal	$4,500

Cost of Retaining the Cut-Off Procedure

Cost/yr of assistant controller	$45,000
No. of business days/yr	÷ 260
Cost/day of assistant controller	$173.08
Days/yr reviewing cut-off	× 6
Cost/yr to review cut-off	$ 1,038

The one-time cost of installing a computer terminal in the receiving area will be offset in about 4.5 years by the labor savings associated with not reviewing the cut-off. However, the controller may be able to justify the expenditure more readily based on the elimination of a key task during the period-end close, which may allow the company to close its books sooner.

An additional issue is whether allowing a late receipt of materials into the warehouse will affect the company's financial statements. As long as a matching cost is on the books, there will be no effect on cost of goods sold and consequently on the income statement. However, is there an effect on the balance sheet? If the J. P. Murphy Company added $78,000 of materials (the largest cut-off error ever found) to its inventory after the close of the period but included it in its period-end financials along with the related payable, its current ratio would be $2,078,000/$1,078,000, which is a current ratio of 1.93:1. This is a decrease in the current ratio to a more conservative ratio than the 2:1 that existed before. If the company's current ratio had been reversed (0.5:1), the cut-off error would have improved it to 0.54:1. In short, a cut-off error will have no effect on the current ratio if a company's current ratio is exactly 1:1, but a current ratio of less than 1:1 will be improved by adding inventory, whereas a current ratio of greater than 1:1 will be changed for the worse.

In summary, the cost/benefit examples in this section can be used as a basis for actual cost/benefit analyses. The examples are linked, so that the costs associated with eliminating part of the receiving function can also be used in justifying the elimination of the warehouse. Thus, when constructing an actual cost/benefit analysis, a better case can be made for change if several of these changes are combined into one cost justification proposal. Also, it would be more accurate to present costs and benefits over a longer time frame and to include a net present value analysis of the longer-term stream of cash inflows and outflows. Cost justifications tend to be more convincing when longer time periods are shown, since one-time project setup costs can be offset by labor savings that continue to pile up over future periods.

REPORTS

The reports needed for a revised inventory system are all similar but are sorted differently and may require minor changes to data elements depending on the purpose of the report. These reports are used to maintain a perpetual inventory system that permits the controller to quickly close each accounting period and track the accuracy and potential obsolescence of the inventory. The information on inventory reports should include each item's inventory location, item number, description, unit of measure (U/M), and quantity. Some reports may also require unit costs and extended costs. By altering the presentation of these reports according to different sort criteria and layouts, this information can be used for cycle counting, inventory audits, checks for mispriced items and incorrect units of measure, and checks for obsolete items. In this section, a sample report is shown for each application.

Inventory Cycle Counting Report

This report is used by the warehouse staff to count blocks of inventory. It is sorted by inventory location. The report may print a blank line in place of the inventory quantity, thereby forcing the inventory counter to manually fill in the inventory quantity rather than conducting a quick comparison of the quantity listed on the report to the amount in the bin. A manual count tends to be more accurate than a quick comparison. An inventory cycle counting report is shown in Figure 3.8.

Inventory Audit Report

This report is used by the internal audit staff to determine the accuracy of the total inventory. It is identical to the inventory cycle counting report except that inventory items are selected at random, so that the auditor can conduct a broad-based review of the inventory. If the cycle counting report were used for this purpose, the auditor would only review a small portion of the inventory at one time, which may have just been counted by the cycle counter and which may not give an

THE HENDERSON GRAPE DRINK COMPANY
Cycle Counting Report
Date _____

Location	Item No.	Description	U/M	Qty
A-10-C	Q1458	Switch, 120V, 20A	EA	____
A-10-C	U1010	Bolt, Zinc, 3 × ¼	EA	____
A-10-C	M1458	Screw, S/S, 2 × ³/₈	EA	____
A-10-C	M1444	Weld Stud, ³/₈ × ³/₈	EA	____
A-10-D	C1515	Flat Bar, 304, 1 × 3	FT	____
A-10-D	C1342	Square Bar, 316, 2"	FT	____
A-10-D	C1218	Round Bar, 305, 1-½"	FT	____
A-10-D	C1110	Weld Pipe, 316, 7"	FT	____
A-10-E	A2700	Sheet Metal, 316, 7 GA	LB	____
A-10-E	A2710	Sheet Metal, 304, 9 GA	LB	____

FIGURE 3.8   Inventory cycle counting report.

THE HENDERSON GRAPE DRINK COMPANY
Inventory Audit Report
Date _____

Location	Item No.	Description	U/M	Qty
A-08-C	M1471	Screw, Tap, 3 × ¼	EA	____
A-12-D	M1100	Bolt, Hex Head, 2 × ⅛	EA	____
A-17-A	M0900	Bolt, Carriage, 4-½″ × ½	EA	____
B-03-C	R0100	Fire Shield, 8″ × 12-½″	EA	____
B-05-E	R1109	Fire Shield, 4-½″ × 12-⅜″	EA	____
B-12-B	R7621	Fire Shield, 6-¼″ × 10-½″	EA	____
C-07-E	C6721	Flat Bar, 316L, ¼″ × 4″	FT	____
D-04-A	C0991	Square Bar, 304L, 4″	FT	____
D-10-C	C8712	Square Bar, 316, 2″	FT	____
E-08-A	A7720	Rapid Transfer Port, 8″	FT	____

FIGURE 3.9   Inventory audit report.

accurate finding regarding the overall accuracy of the inventory. An inventory audit report is shown in Figure 3.9.

Inventory Valuation Report

This report is used by the controller to review the valuation of inventory items. It includes all the information on the cycle counting and audit reports as well as the unit cost and the extended cost of each item. When it is sorted in descending order of extended cost, the controller can review the most expensive items for accuracy. Usually, a quick comparison of the extended cost to the part description will suffice to reveal any items that have incorrect extended costs. One of the primary reasons for an incorrect extended cost is an incorrect unit of measure, so it is important that the units of measure be included in the report. Also, the controller should occasionally review the less expensive items to see if the reverse has occurred—that very expensive items are being costed at excessively low valuations. An inventory valuation report is shown in Figure 3.10.

Inventory Usage Report

This report is used by the materials department and the controller to pinpoint low-usage items for deletion. It shows all the information in the inventory valuation report plus the last date of use. The report is sorted by date, with the oldest date first. The items with excessively old last-use dates are possibly obsolete items. Items thus noted are subjected to a review by the materials review board to

THE HENDERSON GRAPE DRINK COMPANY
Inventory Valuation Report
September 1, 1997

Location	Item No.	Description	U/M	Qty	Cost ($)	Total Cost ($)
C-04-B	C1180	Square Bar, 316L, 4"	FT	150	62	9,300
E-08-A	A7720	Rapid Transfer Port, 8"	FT	9	972	8,748
D-02-D	U1010	Isolator Shell, 4' × 8'	EA	5	995	4,975
B-03-C	R0100	Fire Shield, 8" × 12-$\frac{1}{2}$"	EA	13	182	2,366
E-07-D	W0009	Switch, 120V, 20A	EA	29	70	2,030
D-04-A	C0991	Square Bar, 304L, 4"	FT	11	60	660
A-03-B	D3425	Flat Bar, 304L, 1" × 4"	FT	42	14	588
F-12-C	J1482	Pipe, Alum, 8"	FT	13	42	546
C-10-C	Q5478	Silicone, White	EA	430	1	430
G-03-A	M1457	Screw, Hex Head, 1 × .5	EA	400	.5	200

FIGURE 3.10   Inventory valuation report.

determine if they can be used or if they should be sold off or scrapped. The controller can use this report as a source of information for the obsolete inventory reserve, since the extended cost of possibly obsolete items is listed here and can be added to derive a total obsolete inventory dollar figure. An inventory usage report is shown in Figure 3.11.

In summary, similar forms can be used for various purposes: inventory cycle counting, auditing, obsolescence reviews, and cost extension reviews. When used

THE HENDERSON GRAPE DRINK COMPANY
Inventory Usage Report
September 1, 1997

Location	Item No.	Description	U/M	Qty	Cost ($)	Total Cost ($)	Last Used
C-04-B	C1180	Square Bar, 316L, 4"	FT	150	62	9,300	01/09/88
A-12-D	M1100	Bolt, Hex Head, 2 × $\frac{1}{8}$	EA	27	1	27	04/07/88
A-10-D	C1218	Round Bar, 305, 1-$\frac{1}{2}$"	FT	58	6	348	05/27/88
A-10-C	M1444	Weld Stud, $\frac{3}{8}$ × $\frac{3}{8}$	EA	992	2	1,984	06/12/88
D-02-D	U1010	Isolator Shell, 4' × 8'	EA	5	995	4,975	07/03/88
D-10-C	C8712	Square Bar, 316, 2"	FT	117	25	2,925	12/28/89
E-07-D	W0009	Switch, 120V, 20A	EA	29	70	2,030	03/30/90
A-03-B	D3425	Flat Bar, 304L, 1" × 4"	FT	42	14	588	04/13/91
C-10-C	Q5478	Silicone, White	EA	430	1	430	04/14/92
A-10-E	A2700	Sheet Metal, 316, 7 GA	LB	782	9	7,038	06/06/92

FIGURE 3.11   Inventory usage report.

together, these reports allow the controller to maintain an accurate perpetual inventory. If the perpetual records are accurate, the controller can avoid time-consuming period-end physical inventories. And if perpetual records are used instead of physical inventories, then the period-end close can be completed more quickly.

TECHNOLOGY ISSUES

Technology has played a key role in speeding the processing of inventory transactions. Most of the technological innovations also automatically exchange data, so that transactions can now be processed even faster than if these items were all stand-alone systems. The following are key technologies affecting the inventory area.

On-Line Inventory Systems

Many companies still record inventory transactions in a card file that must be laboriously updated and cross-checked to ensure accuracy. In addition, the file is not linked to the company's accounting system and must therefore be periodically compiled and summarized into the general ledger. This is a very time-consuming task. With the advent of networked computers, companies of any size can install a perpetual inventory system that allows the user to update the inventory records on-line. These systems eliminate the need to manually compile inventory quantities, assign costs to the quantities, and post the information to the general ledger.

Bar-Coded Transactions

Bar codes provide an error-free and extremely fast way to enter inventory transaction information into a computer system. A bar code system is usually linked to a perpetual inventory system, so that scanned information is immediately posted to the relevant inventory part number. Bar codes use a unique set of bar widths and distances between bars to represent alphanumeric digits. When scanned with a laser, this information is decoded into alphanumeric digits and passed directly into the computer database as though the information had been typed using the computer keyboard. It is nearly impossible to scan a bar code and retrieve incorrect information, so bar codes help to reduce data entry error. Unfortunately, errors can still occur if the wrong digit was originally used to create the bar code. Bar codes can be used to record inventory part numbers, quantities, location codes, and transaction codes. For example, if a warehouse worker were to use bar codes to record a receipt of stock into a specific warehouse location, he could scan the part number on the item and the location code taped to the bin, scan a quantity from a quantity chart, and scan a receipt code from a transaction code chart; no keypunching would be necessary.

Radio Frequency Bar Code Scanners

The trouble with early bar coding was that the item to be scanned had to be brought to a computer terminal, where the scanner was located. By attaching a radio to the scanner, the scanner could be brought to the inventory. Also, when inventory had to be moved to a fixed scanning station, warehouse workers tended to jot down the transactions on paper and manually enter the information later, thereby reintroducing data entry errors or forgotten entries. By moving the scanner to the inventory, there is no incentive to delay entry of the information. Also, radio frequency scanners result in truly up-to-the-minute database accuracy, since transactions (depending on the software used) can be posted to the central database immediately. Some software packages only post on a batch basis, no matter how quickly the information arrives from a radio frequency scanner, which negates compromises the usefulness of the technology.

Robotics

Picking systems are available that will pick inventory from a specified location in the warehouse, enter the transaction in the computer database, and deliver the materials to the appropriate location in the warehouse or the production facility—all automatically. These systems are customized and extremely expensive and, because of the cost, are not used by any but the largest warehouse facilities.

In summary, technology has become a key factor in reducing transaction speed in the inventory area. Without computers, companies would still use cumbersome card-based perpetual inventory systems that are difficult to search, summarize, and adjust.

MEASURING THE SYSTEM

Several measures give some indication of a company's inventory-related performance. However, the reasons why inventory statistics fluctuate must be understood, for bad management practices may underlie an otherwise reasonable statistic.

Inventory Turnover

The most widely used measure of inventory performance is turnover. It measures the manufacturing system's efficiency in using inventory and is derived by dividing the usage factor by the average inventory. For example, the turnover of various inventories would be determined as follows:

- Finished goods: Cost of goods sold ÷ average inventory of finished goods.
- Work-in-process: Cost of goods completed ÷ average inventory of work-in-process.
- Raw materials: Materials placed in process ÷ average inventory of raw materials.
- Supplies: Cost of supplies used ÷ average supply inventory.

The result is the number of turns, usually measured in turns per year. Turnover statistics must be analyzed with caution, for different causes can underlie the same result. A slow turnover can indicate overinvestment in inventories, obsolete stock, or declining sales. However, it may simply mean that a company is stocking up for a large custom job with parts that have long lead times. A very high turnover can indicate improved utilization through conversion to a JIT or an MRP system, or it may be caused by keeping excessively small stocks on hand, resulting in lost sales or increased costs due to fractional buying. Many industries publish their average turnover ratios, so benchmark information may be available.

The purpose of business is turning a profit, not turning inventory. Evaluating a company's performance based on just the turnover statistic is not wise without more detailed information. If turnover is used to evaluate the performance of a new manufacturing system, such as MRP or JIT, then it is useful. If it is used to compare performance between accounting periods, it is useful as an indicator of underlying problems or improvements that must be researched further to determine the exact causes of any changes in the statistic.

Inventory Accuracy

The accuracy of the inventory database is measured as total number of errors discovered ÷ total number of inventory items reviewed. For purposes of this calculation, there can be only one error per inventory item (otherwise there might be a negative accuracy percentage).

Errors in inventory can be caused by incorrect quantities, locations, and units of measure for an inventory item. A high inventory accuracy number means that the information in the inventory database can be relied on to yield accurate information about available inventory, which can be used for purchasing and production control decisions. The inventory accuracy statistic can be skewed if the items counted are in the most recently cycle-counted area of the inventory; instead, the count should randomly review the entire inventory.

Bill of Materials Accuracy

The accuracy of the bill of materials database is measured as total number of errors discovered ÷ total number of bills of materials reviewed. For purposes of this calculation, there can be only one error per bill of materials.

Errors in bills of materials can be caused by incorrect part quantities or units of measure and the level of the bill at which a part is included. A high bill of materials accuracy number means that the information in the database can be relied on to yield accurate information about the parts content of items to be produced, which can be used for purchasing and production control decisions.

Inventory Support Cost

This statistic is used to determine how much the company is spending to maintain its inventory. An excessively high ratio may trigger a cut-back in warehouse-related expenses or a decision to adopt a JIT or an MRP system, both of which can reduce inventory levels and support costs. The calculation is as follows:

Cost of warehouse salaries and related benefits

+ Cost of warehouse depreciation

+ Cost of interest on money invested in inventory

+ Cost to rent warehouse space

+ Cost of insurance on inventory

+ Cost of obsolete inventory

= Cost of total inventory

In summary, a few key statistics can give management a clear idea of the velocity of inventory, the accuracy of databases that affect inventory, and the cost to maintain inventory. This information can then be used to make decisions to improve inventory utilization, database accuracy, or the cost-effectiveness of maintaining inventory.

IMPLEMENTATION CONSIDERATIONS

This section offers advice on implementing the innovations presented in this chapter and notes the most common pitfalls to avoid. Some of the innovations are mutually exclusive (e.g., you don't need a perpetual inventory if you have decided to eliminate the warehouse), but it seems reasonable to include in this section information regarding all of them.

The recommendation requiring the most detailed implementation schedule is the *perpetual inventory system.* Note that a perpetual inventory system should be implemented using a correct sequence of steps. Not doing so can seriously prolong the time and expense needed to implement the system, so it is important to carry out the implementation in the following sequence:

1. Select and install inventory-tracking software.
2. Test inventory-tracking software.
3. Train the warehouse staff.
4. Revise rack layout.
5. Create rack locations.
6. Lock the warehouse.
7. Consolidate parts.
8. Assign part numbers.
9. Verify units of measure.
10. Pack the parts.
11. Count items.
12. Enter data into computer.
13. Quick-check the data.
14. Initiate cycle counts.
15. Initiate inventory audits.
16. Post results.
17. Reward staff.

A flowchart of the perpetual inventory installation process is shown in Figure 3.12.

Select and Install Inventory-Tracking Software

A perpetual inventory system can be installed on a card-based manual system, but this information cannot be disseminated to the company through a computer system, and therefore has less effect on overall corporate efficiency. Inventory-tracking software should do the following.

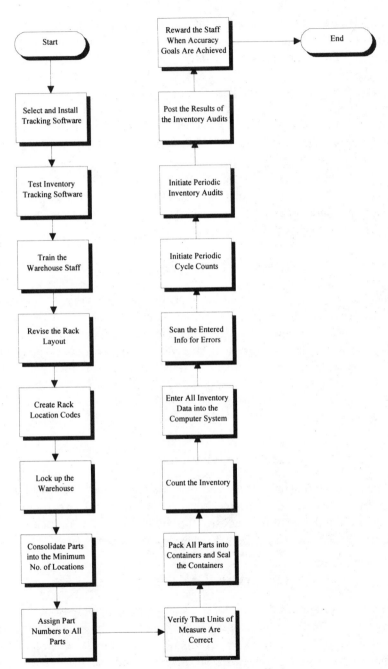

FIGURE 3.12   Perpetual inventory installation process.

Track Transactions.   An important use of a perpetual inventory system is the ability to record the frequency of product usage, which allows the materials manager to increase or reduce selected inventory quantities.

Update Records Immediately.   The perpetual inventory data must always be up-to-date, because production planners must know what is in stock and because cycle counters must have access to accurate data. Batch updating of records is not acceptable.

Report Inventory Records by Location.   Cycle counters need inventory records sorted by location in order to most efficiently count the inventory. Also, the company's internal auditors can spot-check the inventory more efficiently if the records can be sorted by bin location.

Test Inventory-Tracking Software

Create a set of typical records in the new software, and perform a series of transactions to ensure that the software functions properly. At a minimum, these transactions should include issues, receipts, and special adjustments based on inventory counts. In addition, create a large number of records and perform the transactions again, to see if the response time of the system drops significantly. If the software appears to function properly, continue to the next step. Otherwise, fix the problems with the software supplier's assistance, or acquire a different software package.

Train the Warehouse Staff

The warehouse staff should receive software training immediately prior to using the system, so that they do not forget how to operate the software. Enter a set of test records into the software, and have the staff simulate all common inventory transactions, such as receipts, picks, and cycle count adjustments.

Revise Rack Layout

It is much easier to move racks prior to installing a perpetual inventory system, because no inventory locations must be changed on the computer system. Create aisles that are wide enough for forklift operation, and cluster small parts racks together for easier parts picking. Also, if there is a small parts counter that issues parts to the production area, small parts racks should be clustered near the counter. Separate racks should be designated for damaged goods and customer-owned inventory.

Create Rack Locations

A typical rack location is, for example, A-01-B-01. The meaning of this location code is as follows:

A   Aisle A
01   Rack 1
B   Level B (numbered from bottom to top)
01   Partition 1 (optional—subsection of a rack)

As one moves down an aisle, the rack numbers should progress in ascending sequence, with the odd rack numbers on the left and the even numbers on the right. This layout allows an inventory picker to move down the center of the aisle, efficiently pulling items based on sequential location codes.

Lock the Warehouse

One of the main causes of record inaccuracy is removal of items by staff from outside the warehouse. To stop such removal, all entrances to the warehouse must be locked. Only warehouse personnel should be allowed access to the warehouse. All other personnel entering the warehouse should be accompanied by a member of the warehouse staff to prevent the removal of inventory.

Consolidate Parts

To reduce the labor of counting the same item in multiple locations, group common parts in one location. A very experienced production employee should do this, since many parts are similar while still being different parts (e.g., screws that have identical dimensions but are made of different metals); an inexperienced staff person might cluster parts that should be categorized differently.

Assign Part Numbers

Have several experienced personnel verify all part numbers. A mislabeled part is no better than a missing part, since the computer database will not show that it exists. Mislabeled parts also affect the inventory cost; for example, a mislabeled engine is more expensive than the item represented by its incorrect part number, which identifies it as a spark plug.

Verify Units of Measure

Have several experienced employees verify all units of measure. Unless the software allows multiple units of measure, the entire organization must adhere to one unit of measure for each item. For example, the warehouse may desire tape to be counted in rolls, but the engineering department prefers to create bills of materials with tape measured in inches instead of fractions of rolls. This can be a source of friction between departments, so the ability to change the unit of measure may have to be reserved to holders of certain passwords in the software.

Pack the Parts

Pack parts into containers, seal the containers, and label them with the part numbers, units of measure and total quantity stored inside. Leave a few parts free for ready use. Only open containers when additional stock is needed. This method allows cycle counters to rapidly verify inventory balances.

Count Items

Count items when there is little activity in the warehouse, such as during a weekend. Elaborate cross-checking of the counts, as would be done during a year-end physical inventory, is not necessary. It is more important to have the perpetual inventory operational before warehouse activity increases again; any errors in the data will quickly be detected during cycle counts and flushed out of the database. The counts must include part number, location, unit of measure, and quantity.

Enter Data into Computer

Have an experienced data entry person input the location, part number, and quantity into the computer (the unit of measure should already be in the computer, as part of the master record for each part number). Once the details have been entered, another person should cross-check the entered data against the original data for errors.

Quick-Check the Data

Scan the data for errors. If all part numbers have the same number of digits, look for items that are too long or too short. Review location codes to see if inventory is stored in nonexistent racks. Look for units of measure that match the part being described. For example, is it logical to have a pint of steel in stock? Also, if item costs are available, print a list of extended costs. Excessive costs typically point to incorrect units of measure. For example, a cost of $1 per box of nails will become $500 in the inventory report if nails are listed in eaches.

Initiate Cycle Counts

Print out a portion of the inventory list, sorted by location. The sort by location is important, since it keeps the inventory counter from wasting too much time walking around the warehouse, looking for widely separated items. Using a form like the one in Figure 3.8, have staff count blocks of the inventory on a continual basis. They should look for accurate part numbers, units of measure, locations, and quantities. The counts can concentrate on high-value or high-use items, but the entire stock should be reviewed regularly. The most important part of this activity is to examine why mistakes occur. If a cycle counter finds an error, the cause of the error must be investigated and corrected, so that the mistake will not occur again.

Initiate Inventory Audits

The inventory should be audited frequently, perhaps as often as once a week. This allows the controller to track inventory accuracy and initiate changes if the accuracy drops below acceptable levels. In addition, frequent audits are an indirect means of telling the staff that inventory accuracy is important and must be maintained. The minimum acceptable accuracy level is 95%, with an error being a mistaken part number, unit of measure, quantity, or location. This accuracy level is needed to ensure accurate inventory costing as well as to assist the materials department in planning future purchases.

In addition, a tolerance level should be established when calculating inventory accuracy. For example, if the perpetual inventory record of a box of screws yields a quantity of 100 and the actual count reveals a quantity of 105, then the record is accurate if the tolerance is 5% but inaccurate if the tolerance is 1%. The maximum tolerance should be 5%, and this figure could be reduced for high-value or high-use items.

An inventory audit form, like the one shown in Figure 3.9, is sorted by location (to keep the auditor from roaming the warehouse, looking for widely separated locations), and the items are computer-selected at random so that the auditor will, over time, review a broad range of items. Also, the computer can be programmed to select a substantial proportion of high-use or high-value items (known as A inventory) rather than low-turnover or low-value goods (known as B or C inventory). Items in the A classification are usually counted more frequently (in more successive counts) than those in the other classifications. Some C items may be counted only rarely.

Auditors consider it best to print a report that does not list any perpetual inventory quantities, thereby forcing the auditor to fill in a manual count on the report and later compare it to a computer-generated report of perpetual inventory quantities. This keeps the auditor from conducting a "lazy" audit and simply checking off quantities on the audit report if counted quantities appear to be approximately correct.

Post Results

Inventory accuracy is a team project, and the warehouse staff feels more involved if the audit results are posted against the results of previous audits. An unusual aspect of inventory accuracy that the controller should be aware of is that inventories tend to become more inaccurate when management pays close attention to eliminating obsolete inventory (which is usually highlighted by a perpetual inventory system that reports on the quantities of inventory used). The reason for this is that obsolete inventory is the most reliable part of an inventory accuracy audit; it never moves, so it is always accurate. When obsolete materials are removed from the inventory, the remaining inventory moves much more quickly in and out of the warehouse, which leads to greater opportunities for transaction errors and thus to a more inaccurate inventory.

Reward Staff

Accurate inventories save a company thousands of dollars in many ways. Therefore, it is cost-effective to motivate the staff to maintain and improve inventory accuracy with periodic bonuses based on reaching higher levels of accuracy with tighter tolerances.

The previous steps dealt with the basic implementation of an accurate perpetual inventory system. However, there are several special cases that require additional steps.

Customer-Owned Inventory

Customer-owned inventory cannot be valued, since the company does not own it. There are different solutions for different companies. For example, one can avoid assigning a cost to the part, assign several part numbers to a part based on who owns it, or segregate the materials in an uncounted area. Care should be taken when assigning several part numbers to the same part, for engineering drawings and bills of materials usually list only one part number for a part.

Consignment Inventory

One technique used by materials departments to improve the production process is to turn over some items to suppliers, who then have title to their own inventory in the production area. Because it is owned by suppliers, it should not be costed. To avoid incorrect costing, this consignment inventory should be stored in clearly marked areas and should not appear in the inventory database.

Materials at Supplier Locations

Company-owned materials are sometimes kept at supplier or customer locations. These items can constitute a large unseen part of the inventory and can easily escape an otherwise rigorous inventory-tracking system. It is the responsibility of the materials department to track this inventory. Track these items by using a special location code for the off-site location, and verify the item quantities with the customer or supplier as part of the cycle count and periodic audit process.

Floor Stock

Floor stock is defined as the fasteners kept on the shop floor to assemble products. These are typically kept in uncounted bins and replenished from the warehouse as the bins empty. The easiest treatment of this material on the computer system is to avoid it. Floor stock is generally not expensive and has no important effect on the financial statements if they are expensed rather than capitalized into inventory. Also, the cost to count floor stock may not be worth the additional level of accuracy in the perpetual inventory.

Another approach to floor stock is to return as much of it as possible to the warehouse. A close review of floor stock turnover typically reveals that some of it turns slowly. If so, those items can be returned to stock and later requisitioned back to the shop floor as needed. This technique reduces the amount of uncounted floor stock.

Using actual costs to value the inventory depends on linking the company's purchase order cost database to its inventory quantity cost database as well as on programming a costing scheme that matches its method of valuing inventory (e.g., FIFO, LIFO, average cost). Since this may be a custom programming effort, not one that is available in off-the-shelf software modules, it may be costly and time-consuming. The problem is exacerbated if the purchasing and accounting databases are not computerized to begin with, or are located in separate, free-standing computer systems. If so, then conversion to a computerized database is required, as well as linking of the computer hardware with a network so the appropriate information can be transferred between databases.

Requiring purchase orders to receive an item brings considerable rigor to the receiving function. Since many receiving departments may be used to accepting whatever comes in through the receiving bay, this will be quite a shock to them and may require multiple reviews by the internal auditing department to verify that the new rule is being followed. Also, this may increase the work load of the purchasing department, which may find that it is creating many more purchase orders than before—this may require extra staff (probably at the clerical level) to ensure that purchase orders are printed for all materials to be received.

Eliminating excess materials requires the cooperation of the manufacturing and engineering staffs, who must decide what parts can be eliminated. The trouble is that the manufacturing staff probably does not want to delete any items, since they are rewarded for shipping products on time and feel they might have a sudden need for items that everyone else considers to be obsolete. As a result, winnowing inventory can create some surprisingly hard feelings on the production floor. One solution is to move all floor stock onto the production floor (always favored by the production staff, since they no longer have to sign out floor stock from the warehouse). The other option is to reward the production staff based on inventory turnover—the higher the turnover, the lower the inventory. Also, the purchasing staff must be involved, since they will be responsible for returning the materials to suppliers or selling the materials to resellers.

When using *bills of materials,* the controller should be aware that these must be kept accurate, since they are likely to become less accurate over time as changes are made to the product without being reflected in the bill of materials record. The controller should ask the internal auditing department to periodically review the accuracy of bills of materials, checking for such items as units of measure, part quantities, and part descriptions. To help keep the bills accurate, the controller should implement procedures to report on excess items being issued from the warehouse to the production floor, since this indicates that the bills of materials are missing some components. Another procedure will be needed to report on items being returned to the warehouse from the production floor, since this indicates that there are too many parts listed on the bill of materials. In addition, forms should be made available to the production staff that allow them to list part shortages or excesses, or items that do not fit well. All this information must go to the engineering department, which may require additional staff to update the bills of materials.

The implementation of *supplier certifications* involves sending an experienced engineer and purchasing person to suppliers to review their operations; since experience is very important, the company should hire replacements for these people if they already hold key positions in the company. Supplier certification may take much longer than expected, because some suppliers will take a very long time to change their processes sufficiently to become certified; also, some suppliers may refuse to cooperate, requiring a search for an entirely new supplier.

Paying suppliers based on production records has several implementation issues. First, bills of materials must be exceptionally accurate, so that correct payments can be made to suppliers. Second, suppliers will require a great deal of persuasion to adopt this technique, since they will complain that they will not be paid for items damaged during the production process and for delivery overages. The company must answer this concern by tracking parts that are damaged during the production process, so that the supplier can receive full credit for the value of the delivery. Alternatively, if the supplier delivers too many parts, payment will not be made on the overage—the whole point of bypassing the receiving function and delivering parts directly to the production process is to eliminate inventories.

In summary, the implementation of nearly every recommendation in this chapter requires considerable additional staff time and sometimes the hiring of new staff. To implement a physical inventory, experienced staff must be transferred to the job for several months. Programming time will be needed to match actual costs to inventory quantities. The purchasing department's work load will go up if purchase orders are required for all received materials. Eliminating warehouse materials requires a large time investment by the staffs of several departments. Maintaining accurate bills of materials requires extra reporting time by the warehouse staff and record maintenance time by the engineering department. Supplier certification systems frequently involve hiring extra staff to review suppliers. When implementing these changes, the controller should budget for extra costs to ensure that the labor is available to complete the projects.

SUMMARY

It is of great importance to speed up the processing of inventory transactions, yielding an inventory database that contains more up-to-date information. When the inventory information is up-to-date, more employees will rely on the information in the database to make decisions. If more people are using the inventory information to make decisions, the information must be as accurate as possible to prevent incorrect decisions from occurring. This requires the use of inventory cycle counts, audits, and costing tests to ensure that the information stays accurate. If all the techniques described in this chapter are used, then the controller can not only avoid period-end inventory counts and close each month sooner but also provide more accurate information to management than would be the case with a physical inventory.

REFERENCES

Whittington, O. Ray, and Kurt Pany. *Principles of Auditing*. Chicago: Irwin, 1995.

Willson, James D., Janice M. Roehl-Anderson, and Steven Bragg. *Controllership*. 5th ed. New York: Wiley, 1995.

4

ACCOUNTS PAYABLE

The accounts payable function is heavily laden with a blizzard of transactions arriving in the accounting department from all directions: receiving transactions from the receiving department, purchase orders and requisitions from the purchasing department, and invoices from suppliers. The accounting staff must reconcile all these incoming transactions, which frequently requires a substantial amount of research. In addition, despite the evidence of at least three different preexisting transactions regarding a purchase, the accounting department sends out the supplier invoice for approval! This sequence of transactions may take a whole month to process (no wonder most companies prefer 30-day payment terms—they can't pay any faster). In this chapter, we review the existing purchasing and payables system, and ways to reduce processing time while still maintaining control over the process.

CURRENT SYSTEM

A typical payables transaction involves many approval steps, takes an inordinate amount of time, and crosses departmental lines regularly. These factors result in perhaps the most confused paperwork flow of any transaction with which the controller must deal.

A purchasing request can begin in any department when an employee fills out a purchase requisition for an item, gets it signed by a manager, and brings it to the purchasing department. That department then adds more information to the requisition, such as the account number to be charged and the supplier to be used. This last item may require considerable time if the purchasing department puts all items over a certain price level out to bid. After the requisition has been completely filled out, a copy of the requisition is filed, another copy goes back to the requester, and another copy goes to the accounting department. Thus, the transaction has already involved three pieces of paper.

The purchasing department then creates a purchase order and has it signed by a manager. These documents are numerically sequenced, and the purchase order stock is usually locked away when not in use. Copies of the purchase order go to the supplier as well as to the accounting and receiving departments, and another copy is filed in the purchasing department (usually with a copy of the requisition). The number of pieces of paper involved has now grown to seven.

Once the ordered part has arrived, a receiving clerk inspects it and marks down the receipt in a receiving log. This log provides evidence that the item has been received, and it is needed by the accounting department as backup for paying the supplier; therefore, a copy of the log is sent to the accounting department. The number of pieces of paper involved has now grown to nine.

The supplier sends the company an invoice for the item just received. The payables clerk compares the quantity listed on the invoice to that on the receiving report and purchase order to ensure that payment is being made only for the item ordered and received. The prices, discounts, and terms of shipment are also reviewed to ensure that no overpayments are made. The purchase order, receiving document, and invoice frequently do not match and must be reconciled. The payables clerk becomes an investigator, checking with the payables department for incorrect terms and prices, the receiving department for the amount received, the shipping company for evidence of shipment, and (most often) with management to see whose in-box currently contains the invoice that may require multiple layers of management approval. In addition, if monthly statements are received from suppliers, they are reconciled against in-house supplier records, and the supplier is contacted if there are any discrepancies. If statements are not received from suppliers, then the only warning the payables clerk receives regarding missing payment information is an angry call from a supplier receivables clerk, wondering where payment is for an item shipped several months ago.

In addition, one accounting systems study[1] claims that the average company has between three and four payables systems per billion dollars of revenue. Larger firms may not even have systems with a common payables database, which can be quite irritating for suppliers who are trying to track down late payments.

The production area is not the only place in a company where one may find work-in-process (WIP). The accounts payable version of WIP is the (sometimes large) set of invoices from suppliers that have not yet been matched to any receiving or purchasing documents, or conversely, reports for which no matching supplier invoice has been found. In either case, it is difficult to determine the company's actual liability, since not enough information exists to enter a cost

---

[1] Hackett Group, *AICPA/THG Benchmark Study: Results Update and Analysis,* 1994.

into the company's general ledger (and supporting records), which may materially alter the company's financial statements.

After all documents for a purchase have been reconciled, the payables clerk authorizes payment to the supplier. A check is printed and mailed. All the documents are stapled together, along with a copy of the check (another document), and filed away. A stripped-down version of this payables process, excluding management approvals and document filing, is shown in Figure 4.1. The flowchart is modified in Figure 4.2 by adding the management approval steps and the time typically required to obtain approvals. The time required to obtain approvals is far in excess of the time required to actually process the paperwork for the payables and create checks for the suppliers.

One of the biggest problems with the typical accounts payable reconciliation process is that the accounting clerk is often deluged with conflicting information from many different sources. This problem is illustrated in Figure 4.3. For example, the incidental information on the supplier invoice may not match the information on the purchase order—the tax rate or shipping and handling prices may vary. The purchase order may vary from the purchase requisition because the wrong part was ordered. The purchase order quantity frequently varies from the amount received (and invoiced). The received amount is frequently miscounted or not recorded in the receiving log at all. If damaged goods are returned or there are overages, the supplier must also be debited or credited for these variances. Finally, there may be a legal contract associated with the payment that must be consulted for such items as scheduled price increases and methods of delivery. In short, the payables clerk faces a monumental task that must be performed for even the smallest, most insignificant items; in fact, cheaper parts tend not to receive as much attention from the purchasing and receiving personnel as expensive items, so more reconciliation work may have to be done for minor parts.

Another drain on accounting staff time is the processing of employee expense reports. As shown in Figure 4.4, the employee must fill out an expense report, attach receipts for the larger items, and send it to the accounting department, where it is reviewed for compliance with company expense reimbursement policies. The report is then sent to the employee's supervisor for further review and approval, and then back to the accounting department for payment. This process consumes a lot of time, and the transfer of documents between several departments leads to the possibility of the employee's not being reimbursed in time to pay for the expenses within the payment time limit of his or her credit card.

Finally, payables documents are very likely to become lost, because they are transferred across departmental boundaries so frequently. As shown in Figure 4.5, departmental transfers occur for all management approvals, purchase requisitions, purchase orders, and receiving logs. Every time such a transfer occurs, it is possible that a document will be delayed or lost. Given the large

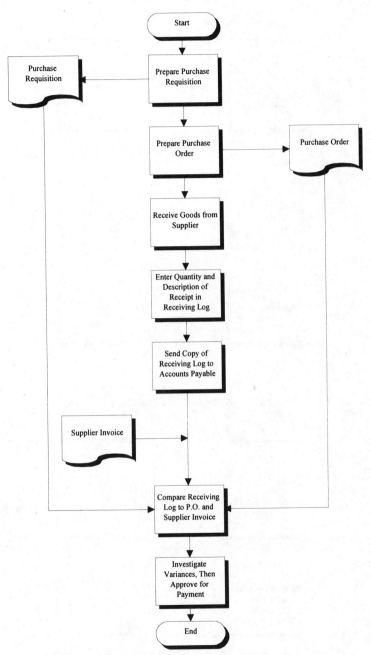

FIGURE 4.1   Typical accounts payable process.

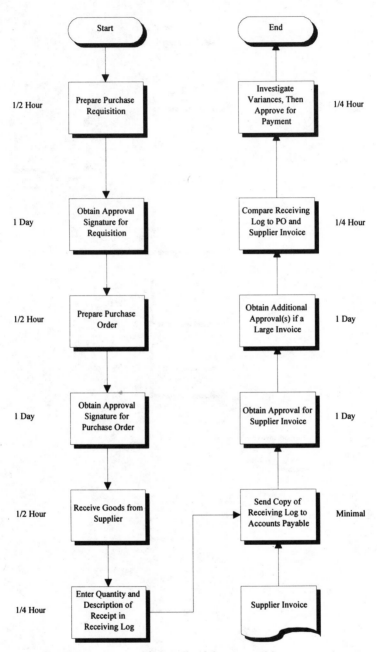

FIGURE 4.2   Approvals required for a payables transaction.

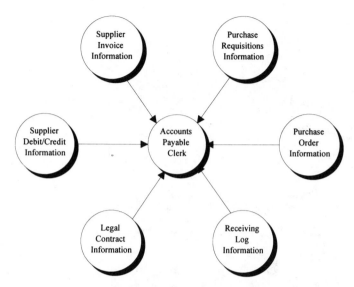

FIGURE 4.3   Multiple sources of accounts payable information.

number of transfers required for the payables process, it is likely that this will happen.

The move and wait times in the traditional payables process greatly slow it down, and every time something is moved, the paperwork may be lost or misinterpreted. The value-added analysis shown in Table 4.1 lists each step in the purchasing and payables process and the time required to complete it. A value-added item is considered to be one that brings the purchasing or payables transaction closer to conclusion.

Table 4.2 shows that only four of the steps bring the purchasing and payables transaction closer to conclusion; the remaining steps are related to moving paperwork from person to person, getting approvals, or making copies for department filing. In terms of time required, the value-added steps can be concluded in 1.7 hours, while the moving, waiting, and non-value-added portions of the transaction take up about two business days. In short, the actions needed to conclude the transaction are only a small part of the total process.

In summary, a typical accounts payable system requires considerable reconciliation work by the accounts payable staff to determine the amount to pay a supplier, based on sometimes conflicting information from a large number of separately maintained databases. This results in a long processing time and a high clerical cost per transaction. The following section reviews several ways to reduce the number of databases and the amount of reconciliation work, resulting in faster completion of the payables process.

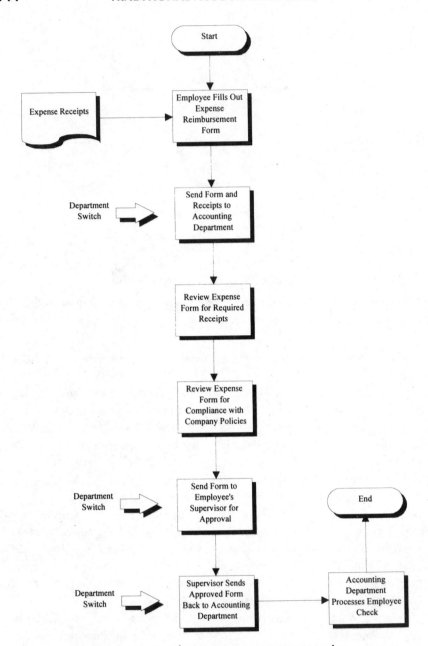

FIGURE 4.4   Employee expense report processing.

	Accounting	Receiving	Purchasing	Legal	Management	Anyone
Purchase Requisition	✓		✓		✓	✓
Purchase Order	✓	✓	✓		✓	
Invoice	✓				✓	
Receiving Log	✓	✓				
Supplier Debits/Credits	✓	✓				
Contract Information	✓		✓	✓		

FIGURE 4.5   Movement of accounts payable paperwork across departments.

TABLE 4.1   Purchasing and Payables Value-Added Analysis

Step	Activity	Time Required (Minutes)	Type of Activity
1	Employee prepares purchase requisition.	5	Non-value-added
2	Employee moves to supervisor.	1	Move
3	Supervisor signs requisition.	1	Non-value-added
4	Employee moves back to office.	1	Move
5	Employee files copy of requisition.	1	Non-value-added
6	Employee moves to mailbox area.	1	Move
7	Employee leaves requisition in mailbox of purchasing department.	1	Non-value-added
8	Wait time before purchasing department picks up mail.	120	Wait
9	Purchasing person brings requisition back to purchasing department.	1	Move
10	Purchasing person obtains order information.	30	Value-added
11	Purchasing person completes a purchase order.	10	Value-added
12	Purchasing person moves to authorized PO signer.	1	Move
13	Authorized signer reviews and signs the purchase order.	1	Non-value-added
14	Purchasing person moves back to office.	1	Move
15	Purchasing person files copy of purchase order.	1	Non-value-added
16	Purchasing person moves to mailbox area.	1	Move
17	Purchasing person mails copy of the purchase order.	1	Non-value-added
18	Purchasing person leaves copy of purchase order in requester's mailbox.	1	Non-value-added

(Continued)

TABLE 4.1 *(Continued)*

Step	Activity	Time Required (Minutes)	Type of Activity
	[Parts are received from supplier.]		
19	Receiving person enters receipt in receiving log.	5	Non-value-added
20	Receiving person moves to mailbox area.	1	Move
21	Receiving person leaves bill of lading in mailbox of accounting department.	1	Non-value-added
22	Wait time before accounting person retrieves mail.	120	Wait
23	Accounting person brings bill of lading back to the accounting department.	1	Move
24	Accounting person files bill of lading.	1	Non-value-added
	[Supplier invoice arrives.]		
25	Mail room person leaves invoice in accounting department mailbox.	1	Non-value-added
26	Wait time before accounting person retrieves mail.	120	Wait
27	Accounting person brings invoice back to the accounting department.	1	Move
28	Accounting person logs invoice into computer system.	5	Non-value-added
29	Accounting person takes invoice back to mailbox area.	1	Move
30	Accounting person leaves invoice in mailbox of authorized approver.	1	Non-value-added
31	Wait time until authorized approver retrieves mail.	120	Wait
32	Authorized approver signs invoice and places in mailbox of accounting department.	2	Non-value-added
33	Wait time until accounting person . retrieves invoice	120	Wait
34	Payables clerk retrieves copy of bill of lading.	1	Non-value-added
35	Payables clerk retrieves copy of purchase order.	1	Non-value-added
36	Payables clerk matches purchase order to bill of lading and invoice.	5	Non-value-added
37	Payables clerk files in one bundle the purchase order and bill of lading.	1	Non-value-added
38	Payables clerk moves to mailbox area.	1	Move
39	Payables clerk leaves invoice in mailbox of authorized signer.	1	Non-value-added
40	Wait time until authorized signer retrieves invoice.	120	Wait

TABLE 4.1 *(Continued)*

Step	Activity	Time Required (Minutes)	Type of Activity
41	Authorized signer reviews and signs invoice.	2	Non-value-added
42	Authorized signer leaves approved invoice in accounting department mailbox.	1	Non-value-added
43	Wait time until accounting person retrieves mail.	120	Wait
44	Accounting person brings mail back to accounting department.	1	Move
45	Payables clerk retrieves the purchase order and bill of lading and matches with invoice.	5	Non-value-added
46	Payables clerk enters approval code into computer system for payment by check.	2	Non-value-added
47	Accounting clerk prints checks.	60	Value-added
48	Payables clerk matches checks to backup packet of purchase order, invoice, and bill of lading.	60	Non-value-added
49	Accounting clerk takes checks and packets to authorized check signer.	1	Move
50	Authorized signer reviews packets and signs checks.	10	Non-value-added
51	Accounting clerk takes checks back to accounting department.	1	Move
52	Payables clerk removes backup packets from checks, attaches copy of check to packets, and files packets.	2	Non-value-added
53	Accounting clerk brings checks to mail room.	1	Move
54	Mail room staff mails the checks.	2	Value-added

TABLE 4.2 **Summary of Purchasing and Payables Value-Added Analysis**

Type of Activity	No. of Activities	Percentage Distribution	No. of Hours	Percentage Distribution
Value-added	4	7%	1.70	9%
Wait	7	13	14.00	78
Move	16	30	.30	2
Non-value-added	27	50	2.02	11
Total	54	100%	18.02	100%

REVISED SYSTEM

The typical purchasing and payables system is burdened by an excessive number of approvals, too many forms requiring reconciliation, and a high volume of items requiring processing. This section examines methods for improving and speeding up the purchasing and payables process.

An important step is *reducing the volume of transactions handled.* An analysis of the dollar value of each payables transaction will reveal that a disproportionate amount of labor is devoted to very small purchases that are not worth all the approvals, purchase requisitions, purchase orders, supplier invoices, matching, and payment with individual checks. Instead, the controller can institute the use of corporate purchasing cards. These credit cards allow employees to purchase small-dollar items directly from suppliers. The advantage to the accounting department is that a large number of transactions are reduced to a small number of monthly credit card payments. A typical purchasing and payables transaction costs about $30 to process; by switching to corporate credit cards, the cost per transaction drops to less than $1. Also, by reducing the overall number of payables transactions, the payables staff can concentrate on the remaining transactions, which usually results in fewer transaction mistakes.

Another way to reduce the volume of transactions handled by the accounts payable department is *simplifying the ordering process for commonly purchased items.* For example, most office supplies are ordered through one office supply company; that being the case, it is easy to create an order form for the most commonly ordered materials and post it in the company's supply room. A periodic review of the supply room will reveal which items are running low; those items can be noted on the order form and faxed to the office supply company for rapid refilling. Many office supply companies even deliver the desired materials free of charge, so no transportation arrangements are needed. The same process can be used for manufacturing supplies, although they are not usually ordered from just one supplier. In either case, the ordering process avoids the usual purchasing and payables process involving purchase requisitions, purchase orders, and multiple management approvals. If a corporate purchasing card is used to automatically pay for the supplies, the controller can also avoid matching receivables documentation to supplier invoices. Since supplies involve moderate dollar amounts but very large transaction volumes, the risk of losing large amounts of money is minimal and the volume of payables transactions is noticeably reduced.

Another way to reduce transaction volume is *automating the expense report processing function.* This system allows employees to enter their expense report information directly into a central computer system, which collects information about specific days being reported, expense items requiring receipts, and explanations for missing receipts. Then the system prints out a transmittal slip, to which the employee attaches all receipts needed for the expense report. The employee gives the transmittal slip and attached receipts to the accounting department, which only has to verify that all the receipts listed on the transmittal slip are attached. The electronic expense report is sent over the computer network to the employee's supervisor for approval. After the electronic approval and the receipts have been received, the accounting department issues an electronic funds transfer (EFT) to the employee. Of course, personal identification numbers (PINs) are used to restrict access to designated computer users. This system allows the accounting department to further reduce its work load by shifting the expense report processing function to the computer system. A flowchart of the process is shown in Figure 4.6.

An alternative way to reduce the review work associated with expense reports is *reviewing expense reports only by random audit.* This greatly reduces the detailed examination currently used in nearly all corporations. Though it is likely that employees will be able to charge off some expenses when an expense report is not selected for an audit, the prospect of being audited will act as a deterrent for most employees. To refine the audit selection technique, the controller should act like the Internal Revenue Service—select employees for more frequent audits if random audits find errors on their expense reports, and select expense reports that contain excessively high travel, hotel, meal, or incidental charges. By refining the selection method for random audit, the accounting department can cut back on the labor required to conduct the reviews, while not giving up all control.

Having reduced the overall volume of payables transactions, the controller must now reduce the amount of paperwork associated with each remaining transaction. The first step is *skipping the supplier invoice.* This has multiple benefits of one less piece of paper to handle, one less item requiring management approval, and one less item to include in the document-matching process. The key to this improvement is to pay the supplier directly from the purchase order. Under this revised system, when a buyer in the purchasing department issues a purchase order to a supplier, the buyer also enters the order into a purchasing database. Suppliers then send the goods to the receiving dock. A receiving clerk checks the purchasing database via a computer terminal located at the dock. The database tells the clerk if the received shipment corresponds to an outstanding purchase order in the database; the information the clerk must match includes the part number, unit of measure, quantity, and vendor name. If it does, then the clerk accepts the goods and enters this information in the database. The computer system should incorporate a preset tolerance level for closing purchase orders even if the amount received is off slightly (a typical tolerance level is 5%). The computer will now automatically print a check to the supplier in accordance with the payment terms on the purchase order. If the received items are not listed on the database, they are rejected on the spot and sent back to the supplier. In short,

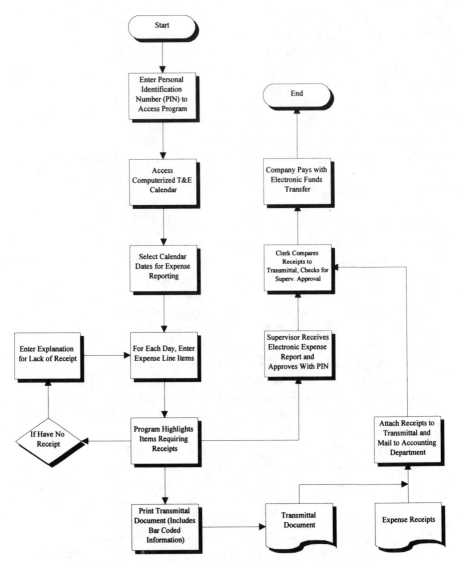

FIGURE 4.6   Automated employee expense report system.

payment authorization has moved to the receiving dock from the accounts payable department. This does not mean that the accounts payable department can eliminate all invoice processing, for there are always quantities of invoices arriving from service providers and for items needed on an emergency basis that do not have a related purchase order. Nonetheless, the quantity of invoice-related work by the payables department drops dramatically. By eliminating the supplier invoice, the speed of processing the payables transaction greatly improves.

Having reduced the overall volume of payables transactions and the need for the supplier invoice, the controller can now shorten the length of the transaction further *eliminating approvals for purchase requisitions*. The bulk of all purchase orders written are for items needed to manufacture products; these purchases are frequently issued automatically by a materials requirements planning (MRP) system or are issued manually based on preapproved bills of materials. These items require no requisition approvals, since the information on which the purchase is based (the bill of materials) has already been approved by management. The next-largest item normally requisitioned is supplies (either for the office or the manufacturing area). With the use of corporate purchasing cards, these transactions no longer require requisitions. This reduces the number of items requiring requisitions to the minimum. For this remaining amount, purchase orders must be generated and approved by management anyway, so the control point is simply shifted to management approval of the purchase order. By eliminating the approval of purchase requisitions, a number of hours (if not days) are removed from the total time needed to process transactions.

The final, and most advanced, step is *paying suppliers based on production records* instead of supplier invoices or (as mentioned earlier) purchase orders. The key item here is to avoid the review by receiving personnel of the incoming shipments from suppliers. Instead, the parts are moved directly to the manufacturing area, where they are immediately used to manufacture the company's product. Then, the amount of each part (as listed on the bill of materials) is multiplied by the number of products completed to arrive at the total number of parts the company has received from the supplier. This unit total is then multiplied by the cost per unit, as noted on the purchase order, thus arriving at a total amount to pay the supplier. Of course, the system must also include reporting for items damaged during production, so that the supplier is paid for these items as well. By using this system, the accounts payable task is reduced to reviewing payments to suppliers that are suggested by the computer system, and printing and mailing out the checks (even this step can be replaced by electronic funds transfers). The role of the internal audit department increases; since the number of process controls are reduced, the auditors must ensure that the remaining controls are strong. Overall, the use of payments based on production records is only possible in a well-run JIT manufacturing system, because the payment scheme will not work unless the following hold:

- Only enough parts are delivered to manufacture the product. If excess parts are delivered, additional accounting is required to determine the number of units for which to pay the supplier; in this case, it would be easier for the controller to pay from purchase order information than from production records.

- Production is rapid. If it takes an inordinate amount of time to manufacture a product, the supplier must wait too long for payment, since payments are based on finished products. This will not be tolerated by suppliers unless increased prices are built into the purchase order, or if parts can be tracked through partially manufactured units.

- Only one supplier supplies each part. It is difficult to pay suppliers when more than one supplier provides a part. When payments are determined based on the parts included in a finished product, there is no way to tell which supplier delivered an included part.

- The bills of materials are totally accurate. If they are not, suppliers will not be paid for the correct quantity of parts delivered. For example, if a bill of materials incorrectly shows too many units of a part being used, the supplier will be overpaid. If the bill of materials contains too few parts, the supplier will be underpaid.

- Product changes are minimal. If the quantity of a part used in the finished product is constantly changed because of design changes, it is very difficult to determine the correct number of parts for which to pay the supplier.

- Bulk purchases are not needed. If bulk purchases are required for key items that can only be procured in volume, because of pricing, packaging, distances traveled, and so on, then they must be paid for by some other means than production records, since not all the materials will be used immediately to complete finished goods.

Another improvement, which is rarely used, is *paying suppliers immediately* rather than waiting for a prenegotiated time period to make the payment. Immediate payment has the advantage of reducing the number of steps involved in the payables process, since there is no storage of payables information until the correct payment date has been reached. It has the additional advantage of reducing the work-in-process of the accounts payable department, since purchasing information can immediately be sent to permanent storage upon payment to the supplier rather than having it wait in the accounting area in temporary storage facilities. Of course, this technique must be offset against lost earnings on cash that will now be given to suppliers instead of the company's investment officer. At a minimum, this method can be used for smaller purchases, so that the company's invested balances are not significantly reduced.

When all the preceding changes have been implemented, the controller will find that there is one more item to fix: *reducing the number of accounts payable systems to one*. Many large companies have more than one system (usually located in different subsidiaries). When a purchasing system has been converted to the high level of automation described in this section, it is very inconvenient to tie the purchasing system into a cluster of different payables systems, since unique programming (as well as a different set of procedures) is required for each payables system in use. Thus, the controller should concentrate on reducing the number of payables systems to one.

Given the number of qualifications just presented, it is evident that paying suppliers based on the amount produced is not possible for many companies. Those companies best able to do so are those with accurate bills of materials, reliable parts deliveries in small quantities, quick production times, and infrequent design changes.

If these conditions cannot be met, the company should pay its suppliers through purchase order documents. The labor involved will still be less and payments to suppliers will still be quicker than if suppliers were paid from their invoices. Figure 4.7 shows a streamlined payables process.

In summary, exceptional improvements in the speed of accounts payable transactions can be realized by paying for small-dollar items with corporate purchasing cards, automating the review of employee expense statements, eliminating supplier invoices, reducing the number of management approval signatures, and paying suppliers based on purchase order information. To accomplish this transformation safely, the controller must pay attention to the new set of controls needed.

CONTROL ISSUES

Many control issues are related to accounts payable, since missing controls can result in major expenditures that may not be authorized by management. For example, if the controller were to eliminate the comparison of purchase orders to invoices by the accounts payable staff, then suppliers could charge excessive rates without anyone noticing. Similarly, eliminating the comparison of supplier invoices to the receiving log would allow suppliers to bill for parts never sent. This section contains examples of what would happen if various payables controls were removed in order to increase the speed of accounts payable transactions, and it suggests possible solutions to the resulting control problems. These examples pertain to the changes suggested previously for a revised payables system.

Stop Requiring Approval of Supplier Invoices

If supplier invoices are approved by management (sometimes multiple times in the case of large-dollar invoices), they present a last-chance review of payments about to be made. If this control is not used, the next most recent chance to catch

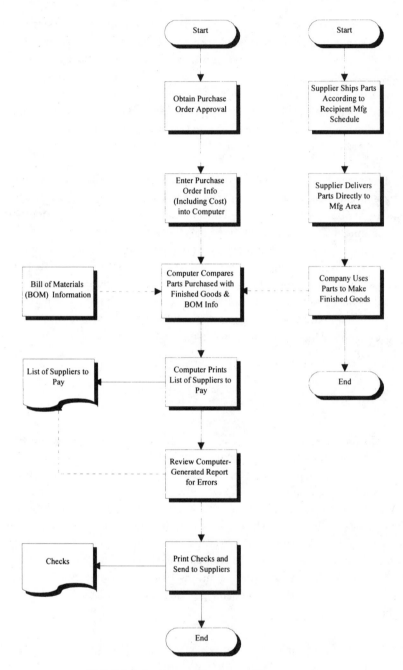

FIGURE 4.7   Streamlined payables process.

an improper payment is the review of purchase orders, while manual signing of checks can also catch errors at that point. For those companies using a signature plate to sign checks, there is no last-minute review of outgoing payments.

Tighten Control over Purchase Orders. If one control is removed, strengthen the remaining ones. The reason why so many approvals are required throughout the payables process is that most accounting systems are so leaky that payments can slip through individual layers of approval but will be caught when the total set of approvals is imposed. The solution to this problem is to tighten control over the key approval point. The company must require that all materials received have an accompanying purchase order number. In order to issue a purchase order, approval must be obtained from management. If management approval is strictly enforced at that level, there is no need for additional approvals downstream from the purchase order point. Enforcing this control point involves rejecting items received that have no purchase order, educating the purchasing staff about this requirement, educating the rest of the company about it, and conducting frequent internal audits of purchase orders to ensure that the correct approvals were obtained.

This control point may not work for service agreements, long-term contracts, and emergency purchases.

Service agreements for such items as janitorial services, lawn care, computer maintenance, and telephone repair services are frequently negotiated by departments other than the purchasing department. Also, service providers do not usually send their staffs into the company through the receiving dock, so they are not logged in. The controller can require a blanket purchase order for each service provider, but it is still nearly impossible for anyone besides the direct contact person within the company to determine if services have been provided. The best control here is to continue to require invoice review by management, since the control point at the purchase order level is comparatively weak.

Many companies negotiate long-term contracts for key supplies; for example, a power plant may have a 20-year contract for coal. In these cases, it is best to create a blanket purchase order that is reviewed periodically to ensure that any changes required by the contract (such as annual price increases) are enforced in the payment system.

If the computer system crashes, the company does not have the purchasing department conduct a competitive bidding process to find a cheap computer repair service—it gets someone who can show up immediately and fix the problem, no matter what the cost. In these cases, the purchasing department is not involved. Instead, a verbal agreement to pay is made by the manager who is trying to fix the problem. In such cases, since the purchase order control point is ignored, invoice approval by management is necessary.

Control Purchase Order Forms.   Management approval of purchase orders is not of much use if purchase order forms can be taken by anyone, typed up, and sent out without management approval. The best control over this problem is to prenumber the purchase orders and secure the purchase order stock. In addition, the purchase order number sequence must be logged in and reviewed for missing purchase order numbers. Finally, the purchase order text should clearly state that the purchase order is invalid without an authorized signature; this prevents a supplier from demanding payment based on a purchase order that has no signature.

Stop Requiring Approval of Purchase Requisitions

If management must sign off on purchase requisitions, managers have an early control over purchases, which is later reinforced by management review of purchase orders before they are issued to suppliers. It also reduces the work load of the purchasing staff, who may otherwise source a part and prepare a purchase order only to find that management disallows the purchase during its review of outgoing purchase orders. If this control were removed, then a leaky purchase order review might let inappropriate purchase orders go out to suppliers.

Tighten Control over Purchase Orders.   Management review of purchase orders was discussed in the previous section.

Systematize the Purchasing of Minor Items.   The volume of purchase requisitions can be greatly reduced by purchasing from a small number of suppliers and creating forms that itemize the most commonly used parts. This eliminates the need for purchase requisitions for most items. These forms are most commonly used for office supplies and production supplies. Since the items involved are usually inexpensive, management approval is not necessary. Instead, the employee simply checks off the items needed and returns the form to the purchasing department, which either orders the materials over the phone or faxes the form to the supplier, who (depending on the size of company orders) may deliver the items to the company. A typical office supplies ordering form is shown in Figure 4.8. In short, by systematizing the purchasing of low-priced, high-volume items, most purchase requisitions can be done away with.

Use Corporate Purchasing Cards.   If the company starts using corporate purchasing cards, the number of purchase requisitions will be reduced, by allowing employees to purchase small-dollar items directly from selected suppliers. Of course, the potential exists for loss of control over purchases; for example, a credit card can be used to purchase inordinate quantities of office or

OFFICE SUPPLIES ORDER FORM
Date _____
Name of Person Ordering _____

Description	Part Description	Supplier Part No.	Quantity
Fasteners			
¼" Alligator Clips	ClipArt ¼"	CA-2500	_____
½" Alligator Clips	ClipArt·½"	CA-5000	_____
Paper			
White Legal	Mead White Long	ME-4320	_____
Yellow Legal	Mead Yellow Long	ME-4321	_____
Staplers			
Stapler, Regular	SwingLine Regular	SW-100A	_____
Stapler, Heavy Duty	SwingLine Heavy	SW-100B	_____
Tape			
¼" Tape	Scotch ¼" Tape	3M-1352	_____
½" Tape	Scotch ½" Tape	3M-1353	_____

FIGURE 4.8   Office supplies order form.

manufacturing supplies, which may then be purloined by the employee making the purchases. However, many corporate purchasing cards have detailed monthly reporting on all expenditures, including a listing of items purchased, where they were purchased, and (less frequently) who purchased them. This reporting allows the controller to conduct an after-the-fact review of purchases and to put a stop to ongoing malfeasance.

Stop Paying Based on Supplier Invoices

If the accounting department pays suppliers based on supplier invoices, the company is assured of having no disputes with suppliers, since payment is based on the information from the database of the supplier. Also, this keeps outside auditors happy, since they confirm payables with selected suppliers and attempt to match the company payables database to the receivables database of the supplier. If this control is not used, differences between the payments made by the company and the payments expected by the supplier will occur with greater frequency, requiring substantial time by the accounting staff to reconcile.

Pay Based on Purchase Orders.   The contract between the company and the supplier is the purchase order, not the invoice from the supplier. As the base-level document, the purchase order with the price noted on it is legally binding.

Thus, the purchase order should be used to pay the supplier. The company can dispense with invoices and match the receiving log to the purchase order, paying the supplier the total of the quantity received times the price on the purchase order. Alternatively, the company can pay the supplier based on th quantity used in the production process (based on the production schedule and the bill of materials) and dispense with the receiving log entirely. Though this system avoids a considerable amount of accounting labor and undoubtedly speeds up the accounting process, several problems may arise with respect to pricing or quantity disputes, supplier debits and credits, shipping charges, and taxes.

From time to time, the supplier will charge a price that varies from the amount on the purchase order. The company's position should be that it will pay according to the price listed on the purchase order; any changes to that price must come from a purchase order change approved in advance by management. Also, the supplier may complain that the company is not paying for the quantity that was shipped. This problem arises when either the supplier or the company miscounted the items shipped. One solution is to rely on the documents of the shipping company as the final arbiter. However, this solution does not work if the materials were not delivered by a third party but instead by the transport of the supplier (increasingly used in JIT environments). If so, the company and the supplier should work to devise a means of packaging that is easily counted; for example, if a box always contains ten units, it is easier to count than a disordered cluster of parts.

There will always be damaged or missing items in shipments that require debits or credits to the supplier. Any shipment overages must be segregated and promptly dealt with by the accounting and purchasing staffs, and the supplier must be notified at once (preferably by fax, to provide a written record) of missing items.

Shipping fees are charged for nearly all deliveries to a company. Since the company is paying from purchase orders, it has no way of knowing what the shipping charge is for each delivery. One way to avoid this problem is to have the shipper charge the company rather than the supplier, thereby avoiding any markups that a supplier may add onto the shipping fee. Another method is to negotiate a standard shipment fee with the supplier in a long-term contract. Then the company can factor the standard charge into its purchase orders and pay the fee based on the purchase order information.

Taxes must often be paid on parts, and the supplier usually provides the tax information on the supplier invoice. If the invoice is being ignored, the company must get the tax information from other sources. In many commercially available accounting packages, tax tables are available by state (and sometimes even by county and city), allowing the company to make tax payments without supplier-provided information.

Stop Matching Purchase Orders to Invoices and the Receiving Log

Purchase orders are matched to invoices and the receiving log in order to verify that the items ordered were received before payment is made to the supplier. Proof of receipt can come not only from the receiving log but also from the supplier, who does not (legally) bill the company until materials have been shipped. In addition, this matching process verifies that the price listed on the invoice is the same as the amount listed on the invoice, so that there is agreement between the supplier and the company regarding the amount to be paid. If this matching control were removed, the company might make payments even if materials were not received and might use an incorrect unit price.

Base Payments on Amounts Produced. If the supplier delivers parts directly to the manufacturing area in the exact quantities needed for production, the production schedule can be used as evidence of materials receipt. The quantity of goods produced are multiplied by the number of parts required for each unit; this part quantity information is listed on the bill of materials. Then the price listed on the purchase order is multiplied by the total number of units used to derive the amount to be paid to the supplier. Of course, any parts scrapped during the manufacturing process will have to be carefully recorded and credited back to the supplier.

Select Expense Reports at Random for Review

Employee expense reports should be reviewed to ensure that only appropriate expenses are being charged to the company. Reviewing a limited number of expense reports allows some employees to charge additional expenses to the company with reduced odds of being caught.

Refine the Expense Report Selection Technique. A few audits will reveal that some employees always report expenses fairly, whereas others always stretch the company rules to report the maximum possible amount. Once patterns emerge, the selection method should concentrate on those employees whose reports continue to reveal reporting problems, and review other employees' reports less often. Also, a cursory review of expense report line items on *all* expense reports will reveal line items that depart noticeably from the amounts to be expected (e.g., for travel, lodging, and meals). When excessive amounts are revealed, just those line items can be audited. In addition, the company will find over time that auditing a certain percentage of expense reports will be cost-effective in terms of the expense errors found versus the labor required to conduct the audits. The company should conduct only the number of audits that brings in maximum expense report reductions for minimum audit labor cost.

Stop Manually Processing Expense Reports

Employee expense reports should be reviewed to ensure that all receipts proving large-dollar purchases have been submitted and that all expenses are reimbursable under company expense policies. Switching to an automated system brings up the problem of access to the system. If someone can file an expense report and approve it with a false personal identification number (PIN), it is possible for excessive reimbursements to take place.

Automatically Assign New PIN Numbers and Access Codes.   A random number generator can easily create new PIN numbers periodically and issue them to employees through the company computer system. However, if a computer access code has also been stolen, the computer is merely issuing a new PIN to someone who has already broken into the system. Consequently, the computer systems administrator must also distribute new system access codes on a regular basis, and not through the computer system; instead, they can be distributed through a more secure medium, such as a note included with an employee's paycheck of EFT payment advice.

Of course, controls are still needed in most corporate environments. The following controls should be maintained, for the potential increase in risk associated with their elimination does not warrant their deletion. Since the recommended changes have deleted nearly every step in the traditional accounts payable process, the controls listed here are mainly related the purchasing and receiving functions as well as to ensuring that payment information reaches the accounting staff so that correct payments can be made.

Select Suppliers

With fewer suppliers, it is easy for one supplier to stop the company's production process with slow deliveries or low-quality goods. This problem can be mitigated by a careful review of each supplier's performance record.

Use Accurate Bills of Materials

When payments are made by multiplying the amount of finished goods produced by the number of parts used in the product, it is very important that the bill of materials (BOM) be correct. To ensure that it is accurate, an engineer must be assigned the task of updating it, access to the BOM file must be severely limited, and processes must be set up so that information about incorrect BOM quantities can be sent to the BOM engineer from all parts of the company.

Account for the Sequence of Purchasing Documents

When the purchase order is the primary means by which both quantities and costs are determined, it is crucial that the form itself be controlled, so that forms cannot be removed and used to illegally purchase materials in the name of the company. This can be achieved by locating purchase order forms in a locked room that also contains the computer printer; this allows forms to remain in the printer, ready for immediate printing, while still maintaining security. A log of sequential form numbers can also be maintained in that locked room; the log should be reviewed regularly to verify that all prenumbered forms are accounted for.

Report Over/Under/Return Conditions to the Accounting Department

With a reduced set of controls covering the accounts payable process, it is extremely important that any changes in the amount received from the amount ordered be carefully noted, so that the amount paid can be changed. If the company elects to keep a receiving department and pay from the purchase order document, then the best approach is to have a computer terminal in the receiving area, on which changes in the received amount can be noted. If no terminal is available, then this information must be manually conveyed to the payables clerk. If the company is paying based on completed production records, then the production staff can report on this information, because the production process will have been interrupted if there is a materials shortage, and there will be excess inventory in the production area if too much material is delivered.

Report Freight Documents to the Accounting Department

Even if freight charges are always paid by the supplier, the company should still retain freight documents, because they provide evidence of receipt. If the company has a dispute with a supplier over whether a shipment was received (and therefore paid for), the shipping documents will be available to provide evidence. When shipments arrive, the freight documents should either be sent to the accounting department or stored securely in the receiving department. Under a JIT delivery system, there may be no shipping documents, since shipments are so frequent and so small that the paperwork burden would be too great for the supplier.

Review Long-Term Contracts

There is a potential for incorrect payments when long-term contracts are used, because fees may change over time and may not be noted by the accounts payable department. Typical contracts of this kind are the building lease and contracts for crucial materials, such as coal for power plants. Thus, the accounting staff should make it a regular practice to review long-term contracts. An alternative is to review long-term contracts as soon as they are signed and create a summary sheet of the most important points, which the controller retains and reviews monthly for changes.

Review Credit Card Documents

Although corporate purchasing cards are not intended for large-dollar purchases, the total volume of low-dollar purchases for which they may be legitimately used is so large that the overall dollar volume involved can be considerable. For that reason, it is wise for the controller to have someone review the credit card statements to determine the types of items purchased, the purchase locations, and the dollar amounts involved. Though the accounting clerk assigned this task may not be knowledgeable regarding specific items purchased, a comparison with dollar amounts from previous months can point to unusual changes in dollar volume between periods. Also, a review by a more knowledgeable person in the purchasing department can supplement the knowledge of the accounting staff.

Use Periodic Audits

The internal audit department can frequently review a number of items surrounding the accounts payable function. Note that internal auditors advise of changes to be made—they do not implement control changes on their own. Thus, the purpose of the following steps is to warn the controller of control problems in selected areas.

Review Duplicate Payments.   This problem can be checked by periodically tracing payments back to their supporting receiving or freight documents.

Review Advance Payments.   It is easy for an accounting clerk to pay a supplier the full amount due, forgetting that an advance may have already been paid. This problem can be controlled by periodically reviewing the details of the supplier advances account and tracing advances back to related receipts from suppliers.

Review Supplier Returns.   Most companies have trouble recording supplier credits for parts returned to suppliers. This problem can be discovered by periodically tracing shipping documents back to the accounting department to ensure that proper credit has been recorded.

Compare the Purchase Order Cost to Paid Cost.   Though this problem arises only when paying from supplier invoices, it is still possible for a company to pay a supplier a different per-unit price than the amount listed on the purchase order. An internal audit comparison of the price on the purchase order to the price paid will reveal this problem.

Compare the Received Quantity to Paid Quantity.   The amount received is the quantity that should be paid for, but many companies pay for the quantity listed on the purchase order. A comparison of the amount paid to the amount received will point out this error.

Compare the Amount Received to Amount Recorded as a Payable.   A major accounting issue is ensuring that accounts payable are not underrecorded. Any problems in this area can be discovered by comparing the quantities received (and their related costs as listed on purchase orders) to the costs recorded in the accounts payable records.

Compare the Amount Requisitioned to the Amount Ordered.   Though rarely a control problem, the amount actually ordered can vary significantly from the quantity requisitioned. This is mainly a problem in a JIT environment, when bulk purchases are not needed and actively discouraged. The audit team can highlight this problem by comparing requisition forms to the related purchase orders.

Review the Sufficiency of Accruals.   Accruals are needed for property taxes, payroll, the pension or 401(k) plan, product warranties, commissions, bonuses, income taxes, and professional fees. Each of these accruals requires a different system to calculate and can be very difficult to record accurately; sometimes the correct accrual is more a matter of opinion than fact. The internal audit team should carefully review the calculations behind each accrual as well as any sudden variations in month-to-month accruals to see if accruals are being used to alter reported earnings.

Review the Payment of Freight and Taxes.   Payments for "incidental" items related to a materials shipment are difficult to track, since the typical company payment system is designed to ensure that the materials received are paid for, not the freight used to get it there or any taxes. Thus, an occasional review of these items by the audit team can be of use.

Review the Accuracy of the Bill of Materials.   For those companies paying suppliers based on the number of parts used in its finished products, it is critical that the company's bills of materials be totally accurate—otherwise, payments to suppliers will be incorrect. The internal audit team can periodically compare bill of materials records to actual usage and note any differences.

Review the Accuracy of the Tax Tables.   Taxes change over time, and the company must pay the correct tax rate. A comparison of actual rates to the amounts listed in the company's tax table will highlight any errors.

Review the Propriety of Purchases Made with Purchasing Cards.   When purchasing cards are used, the potential exists for abuse of the cards by making extraneous purchases. An occasional review of the credit card payment records for the appropriateness of purchases will note any problems in this area.

Review the Sequence of Purchase Order Numbers.   When purchase orders are being used as the primary method for paying suppliers, control over the purchase order forms is critical, since illicit orders can be placed with stolen forms. Thus, a continuing audit review should encompass a cross-check of purchase order storage controls as well as the sequencing of purchase order numbers.

Review the Amount of Service Charges.   One of the most difficult items to track is payment for services, since there is not supporting receiving documentation. To spot any problems in this area, the internal audit team should compare payments to sign-ins by service personnel in the company visitor log, discuss approvals of service provider invoices by the managers who signed the invoices, and compare service provider rates to those charged by other suppliers in the local market.

Compare the Amount of Payments to Long-Term Contracts.   A common problem is for prices paid to gradually diverge from the prices listed in long-term contracts, because no one reviews the contracts periodically. This problem can be highlighted by an internal audit comparison of prices paid to prices agreed upon.

In summary, the very slow accounts payable process can be speeded up more than any other accounting function. This can be accomplished by eliminating a number of the approvals commonly required before a supplier invoice is paid. Another technique is to eliminate the invoice entirely and pay from the purchase order. Also, the volume of small payments can be drastically reduced by using corporate purchasing cards. In addition, highly automated employee expense reimbursement systems can be designed that almost completely eliminate the review of expense reports by the accounting department. Taken together, these changes allow the controller to process payments faster while exerting less effort to do it.

QUALITY ISSUES

Accounts payable transactions are slowed considerably by the introduction of errors into the process. Thus, high-quality accounts payable processing is defined as processing transactions with the smallest number of mistakes. This section lists the most common accounts payable processing errors and notes ways to keep them from happening.

A common payables error is paying for a supplier invoice for which no receiving information exists. The problem is caused when the receiving dock employees neglect to write down the information about a receipt in the receiving log. The accounts payable clerk is the person who must track down the missing information, frequently by working with the receiving department to request proof of shipment from the freight company. To avoid this problem, the best approach is to install a computer terminal at the receiving dock and only allow items to be received if a purchase order for the item is listed in the on-line database. The receiving clerk must then enter the date and quantity of the receipt in the purchase order record. This approach requires the receiving staff to check for a purchase order in the computer database, since the only alternative is to reject the shipment and send it back to the supplier. When there is no alternative between entering the receipt in the computer and rejecting the shipment, there should be no payments to investigate for which there is no receiving information.

A variation on the previous problem is paying a supplier for an amount that was incorrectly counted at the receiving dock. The most common cause of this problem is the packaging of incoming goods in such a manner that they are difficult to count. The best solution is to negotiate with the supplier to prepackage the parts in easily countable containers, such as in units of ten. Another option for precertified suppliers is to have them mark the quantity on the delivered pallet (the most error-free approach is using a bar code), so that the receiving personnel can record the information without conducting an error-prone count at all. This option should only be used with reliable suppliers whose operations have been reviewed by the company's purchasing and engineering staffs and found to meet the company's supplier standards.

Another error is paying a freight company for shipping an item for which there is no receiving records. The most common cause of this problem is loss of the information at the receiving dock by inattentive employees. This problem is solved by incorporating a computer terminal into the receiving process, as previously noted. When the receiving clerk accesses the computer database for purchase order information about the shipment on the receiving dock, the computer can prompt the user for information about the freight carrier. This prompt requires the clerk to input freight information before the record can be entered. By requiring the entry of freight information before the receipt can be processed, all freight information will be entered into the payables system.

Another error is paying for a supplier invoice for which there is no purchase order. There are a number of reasons why this problem occurs. First, employees are giving verbal purchase orders to suppliers, which the accounts payable clerk must now investigate. Second, fraudulent shipments are being made that the company does not want (which is why there are no purchase orders). Third, it may be customary for the providers of services to work without an authorizing purchase order. Unfortunately, the best solution is to issue a companywide policy about the absolute need for purchase orders for all purchases (over a minimum dollar level) and enforce the edict by placing a moratorium on any payments that do not have a purchase order number. The accounts payable staff will then have to investigate each invoice that has no supporting purchase order information and inform employees of the purchase order policy. A second policy should require the issuance of blanket purchase orders to service providers. This extra work will be considerable and will slow down the speed of payables transactions, but it is necessary to halt the issuance of verbal purchase orders, payments for fraudulent receipts, and verbal service contracts. After employees learn about these new policies, the payables staff will find that the number of transgressions declines, though it never totally goes away—after the break in period, the overall work load of the accounts payable department in this respect will decline.

Another error is paying the incorrect periodic amount on a recurring payment. The most common cause of this problem is the incorrect tracking of price changes that are listed in long-term contracts. The best solution is to create a summary of price changes on long-term contracts that is reviewed monthly for changes. This review should be incorporated into the department's monthly activities calendar to ensure that the review is made. In short, incorporating the contract review into the department's periodic calendar of events eliminates the problem.

Another error is losing track of employee expense reports when they are sent to managers for supervisory approval. The cause of this problem is that the expense reports sit in managers' in-boxes without anyone following up to see what has happened to the paperwork. One solution is to enter the expense report information into the accounting database prior to sending it out to supervisors for approval. Then a report should be created that lists expense reports that have been entered into the system but not yet approved for payment. A quick scan of this report will reveal those expense reports that are overdue. A weekly review of this report should be sufficient for timely collection of the expense reports. The accounting manager can then call the tardy supervisors with a reminder notice. Thus, allowing the computer system to remind the user of missing approvals reduces the possibility of payments being overlooked.

Another error is caused by paying suppliers based on inaccurate bill of materials (BOM) information. This error only occurs if payments are made based on production records. If that is the case, then product information is periodically multiplied by each product's BOM to determine how many units for which to compensate the supplier. The causes of the problem are inadequate control over access to the BOM record and lack of control over physical changes to the finished product. Though the solution falls outside the supervisory range of the controller, the effect of the problem hits the controller's area of responsibility; therefore, the solution is noted here. The first step in fixing this problem is to have the internal audit department conduct a periodic review of the accuracy of the BOM, so that management can make further corrections as needed. The *minimum* level of BOM accuracy is 98%. Then a person must be given complete responsibility for the accuracy of all BOMs, with the power to implement all controls needed to keep the BOM records accurate. Some of these BOM controls include minimal access to the BOM record in the computer and reporting from the shop floor and warehouse regarding part quantities that are incorrect. Another critical control is careful review and approval of all proposed changes to a BOM, so that BOM changes are well considered and properly timed. Consequently, control over changes to BOMs and access to BOM information are important ways to improve the accuracy of payments to suppliers.

A considerable source of error is the quality of the goods received from suppliers. If materials are defective, then some or all of a shipment must be returned; this additional paperwork must be noted in the next payment to the supplier. The source of the problem, of course, is the supplier, who did not ship a product that passes the company's quality standards. The only solution is tracking the number of defective parts by supplier, as well as dates received, so the company can determine the percentage of defective parts received from a supplier as well as any trends over time. Unacceptable supplier performance in this area must lead to termination of sales from the supplier. Requiring suppliers to improve their product quality or changing to new suppliers is the only way to reduce the number of transactions involving product returns.

One of the largest sources of information errors is the supplier. The cause of the problem is the suppliers' inability to track the amount or price of product they ship to the company, resulting in inadequate invoice information coming from the supplier; even though this is the fault of the supplier, the error must be investigated by the company's accounts payable personnel. As noted previously, the company should track the quantity of disputes with its various suppliers and terminate those with consistently poor records.

In summary, quality in the accounts payable process is related to a number of items: the quality of supplier information and products, the accuracy of bill of materials information, expense reports sent out for approval, and inaccurately recorded information at the receiving dock. Since nearly all the problems highlighted in this chapter involve problems in other departments (and other companies), the controller must be prepared to improve processes outside of the accounting department in order to improve the quality of the output provided by the accounts payable department.

COST/BENEFIT ANALYSIS

This section describes how to write cost/benefit analyses for corporate purchasing cards, automated expense reimbursements, paying from purchase orders, skipping purchase requisition approvals, and paying from production records. In these examples, expected costs and benefits are as realistic as possible. The justifications are reduced to a one-year time frame in order to reduce the size of the examples; in reality, net present value calculations for five years of costs and benefits are more accurate.

Use Corporate Purchasing Cards

You want to introduce corporate purchasing cards at your company, and need to present a cost/benefit analysis to the purchasing manager. You find that the annual fee of each of 20 corporate purchasing cards is $20. You also conduct an analysis of all company purchases and find that 20% of all purchase orders are for purchases of less than $100, with an average purchase price of $52. Of those purchases, 75% are from five local supply companies, all of which accept credit cards as payment for goods. The company issued 28,000 purchase orders in the previous year, totaling $33,600,000. An activity-based costing analysis has shown that the average purchase order, including all related activities, requires $30 to complete. A local internal auditors trade association has conducted a confidential survey of employee theft using corporate purchasing cards and concluded that improper use increased the cost of items purchased with corporate purchasing cards by 0.5%. Is is worthwhile to use corporate purchasing cards?

Solution.   The following calculations show the cost of using (or not using) corporate purchasing cards.

Cost of Using Corporate Purchasing Cards

Number of corporate purchasing cards	20
Annual fee per card	× $20
Total annual fee for cards	$ 400
No. of purchase orders	28,000
Percentage under $100	× .20
No. of purchase orders under $100	5,600
Percentage of under-$100 POs allowing credit	× .75
No. of under-$100 POs allowing credit	4,200
Average amount/under-$100 PO	× $52
Amount of POs converted to credit cards	$218,400
Expected percentage of improper credit purchases	× .5%
Total amount of improper credit purchases	$ 1,092
Total cost of using corporate purchasing cards	$ 1,492

Cost of Not Using Corporate Purchasing Cards

No. of purchase orders	28,000
Percentage under $100	× .20
No. of purchase orders under $100	5,600
Cost to process a purchase order	× $30
Total cost to process under-$100 POs	$168,000

The analysis reveals that the costs associated with using corporate purchasing cards are minuscule compared to the savings resulting from their use. In fact, based on these calculations, the conversion would pay for itself 113 times over in the first year, which is better than a three-day payback period.[2]

Install Automated Expense Reimbursement System

Giles von Strohe, the CFO of the Near Beer Company, wants to install an automated employee expense reimbursement system. He asks you to explore the costs and savings of such a system and to recommend a course of action. You find that the company's 65 salespeople submit expense reports every week for an average of 40 weeks per year, which is a total of 2,600 expense reports per year. An accounts payable clerk reviews each expense report to see if expenses claimed are in accordance with company policies, matches receipts to the expenses listed on the summary sheet, and resolves discrepancies by calling the salesperson. The

---

[2] This analysis can include a range of possible losses related to employee misuse of the corporate cards, rather than an analysis based on an average rate.

typical expense report review requires 45 minutes. The average payables clerk wage (including benefits) is $10.35. Also, the sales manger reviews each expense report for 5 minutes and usually spends another minute on the phone with the salesperson, asking questions about the expense report. The sales manager's salary is $100,000. Should the system be installed?

Solution.   The analysis is as follows.

Cost of Manual Expense Reimbursement System

No. of expense reports	2,600
Clerical time to review report	× .75 hr
Total clerical hours	1,950 hrs
Cost/hour or clerk	× $10.35
Total cost of clerical review	$20,183
No. of expense reports	2,600
Sales manager time to review report	× .10 hr
Total manager hours	260 hrs
Cost/hour of manager	× $48.08
Total cost of manager review	$12,501
Total cost of expense report review	$32,684

Further investigation reveals that an automated expense reimbursement system cannot be purchased as an off-the-shelf application; instead, it must be programmed in-house. The company's programmer earns $22 per hour. The MIS director estimates that this task will require ten months of the programmer's time, including requirements definition, debugging, training of the sales staff, and creating a user guide. In addition, the 65 salespeople must be trained one-on-one by the programmer. Each training session will take one hour. The average salesperson earns $65,000 ($31.25 per hour).

Cost of Automatic Expense Reimbursement System

Cost/yr of programmer	$45,760
Ten months' work	× .83
Total programmer cost	$37,981
Average cost/hour/salesperson	$31.25
Training time	× 65 hrs
Total training cost	$ 2,031
Total implementation cost	$40,012

In short, your investigation reveals that an automated system will save $32,684 per year on an ongoing basis but will require $40,012 of up-front costs to implement. The payback is 1.2 years. This seems like a reasonable project to recommend.[3]

Pay Suppliers from Purchase Orders Rather Than Invoices

In an effort to cut costs, the CFO of Daisy Baby Foods wants to start paying suppliers from purchase orders rather than from supplier invoices. As the controller, you are assigned the task of creating a cost/benefit analysis. You find that the requirements definition, programming, testing, and documentation work related to the changeover will require the time of three programmers for half a year. The average programmer is paid $37,000. In addition, the mailing and staff costs associated with notifying key suppliers of the changeover will be $25,000. Training time for the staff that will be using the new system will cost about $17,000. You also estimate that two accounts payable clerks will be needed even after the new system is installed, because there will still be a limited number of supplier invoices arriving that do not have related purchase orders; in addition, the clerks must resolve payment disputes with suppliers, which will occur no matter what payment method is used. Despite these issues, the new system should allow the company to reduce the accounts payable department by four positions; the average salary of those positions is $19,500. Is this a cost-effective project?

Solution.   The costs and savings resulting from the project are as follows.

Cost to Implement a Payment System Base on Purchase Orders

Cost/yr of programmer	$37,000
No. of programmers required	× 3
Six months' work	× .50
Total programming cost	$55,500
Mailing and notification cost	$25,000
Training cost	$17,000
Total implementation cost	$97,500

---

[3] A cost/benefit analysis of this project could also include the margin on lost sales, since the sales force is being taken away from sales work to learn how to use the system. This number, however, is not easily quantifiable.

Cost Reductions with New System

Cost/yr of payables clerk	$19,500
Clerical positions eliminated	× 4
Total salary-related savings	$78,000

The cost to install the purchase order-based payment system is $97,500, versus annual savings of $78,000. This is a payback of 1.25 years. Based on such a rapid payback, this project should proceed.[4]

Eliminate Purchase Requisition Approvals

A reengineering team at A.L. Skrupp Corp. has recommended that purchase requisition approvals be eliminated. The controller, Mr. Boldrup, reviews the situation and finds that the average company manager can review, sign, and forward a purchase requisition in 2 minutes. The average time required to find a manager who can sign the requisition is one day. Discussions with a sampling of company managers reveals that they typically reject only 2% of all purchase requisitions sent to them for approval. They are later called upon to sign off on completed purchase orders that contain the same information. The somewhat annoyed purchasing manager informs Mr. Boldrup that it takes 45 minutes of purchasing department time whenever it converts a purchase requisition into a purchase order; if an unapproved requisition is later rejected in purchase order form by a manager, then that is a waste of 45 minutes of purchasing time. The average company buyer earns $18.35 per hour. The average company manager earns $32.25 per hour. The purchasing department receives 82,500 purchase requisitions per year. Should the A.L. Skrupp Corporation eliminate manager sign-offs on purchase requisitions?

Solution.   The following cost analysis reveals the answer.

Cost of Retaining Purchase Requisition Approvals

Cost/hr of manager	$ 32.25
Minutes/hour	÷ 60
Cost/min. of manager	$ .54
Time to review requisition	× 2 min.
Cost/requisitions of management review	$ 1.08
No. of requisition/yr	× 82,500
Total cost of management review	$89,100

---

[4] A more detailed analysis could include the cost of reinforcement training for the staff in future years.

Cost of Eliminating Purchase Requisition
Approvals

No. requisitions approved/yr	82,500
Percentage of requisitions not approved	× 2%
No. of requisitions not approved	1,650
Time to create purchase order	× .75 hr
Total time to create purchase orders	1,237.50 hrs
Cost/hour of buyer	× $18.35
Total cost of creating unneeded POs	$22,708

The cost/benefit analysis reveals that reviewing all purchase requisitions is costing the company a startling amount of management time, which is four times more expensive than the added cost to the purchasing department caused by extra requisitions being converted into purchase orders. An objection to this type of analysis is that the company will not actually realize any cost savings, since the managers will still be paid by the company while the purchasing department will *add* a buyer to handle the extra purchase requisitions being sent to the purchasing department. One response to this objection is that eliminating the signature approval also reduces the requisition processing time by a day. Another response is that managers who are no longer reviewing requisitions can busy themselves with management work instead of clerical work—since this is their proper role, the company may realize benefits in other areas that are caused by increased management attention.

Pay Suppliers from Production Records

The CFO of the Shine Bright Lamp Company is interested in paying suppliers from the information contained in the company's production records rather than from supplier invoices. The CFO's assistant, Mr. Dunwoody, is asked to construct a cost/benefit analysis of the project. He notes that one half of all payments based on supplier invoices could be switched to payments based on production records. This would allow the company to cut its accounts payable department in half. The department has a staff of six, who are paid an average of $19,500 each. In addition, Mr. Dunwoody finds that the programming cost of paying from production records is substantial—four programmers will be required to design, create, and test the needed software over a period of six months. The company's average rate of pay for programmers is $39,000. Also, the 18 key suppliers who will be affected by this change must be visited and informed of the new payment method. The company will send two programmers to the supplier locations to discuss the changes in their systems that will be required as a result of the system change. Those two people will be visiting suppliers full-time for six months. Also, the purchasing manager will visit all 18 companies in advance during a one-month road trip to prepare the suppliers for the change. The purchasing manager earns $70,000 per year. Total travel costs for the programmers and the purchasing manager will be $43,200. Finally, the company's bills of materials (BOMs) are not yet accurate enough, so an engineer must be hired who will review the BOMs on a continuing basis. The salary of the engineer is expected to be $48,000. Should this project be implemented?

Solution.   The following analysis will assist in making the decision.

Cost of Payment from Production Records

Cost/yr of programmer	$ 39,000
No. of programmers	× 6
Six months' work	× .50
Total programming cost	$117,000
Cost/yr of purchasing manager	$ 70,000
One months' work	× 8.3%
Total purchasing manager cost	$ 5,810
Total travel cost	$ 43,200
BOM engineer salary	$ 48,000
Total implementation cost	$214,010

Benefit of Payment from Production Records

Cost/yr of payroll clerk	$19,500
Clerical positions eliminated	× 3
Total savings	$58,500

It is evident from the preceding information that the payback period for this project would be too long for many companies. Therefore, it is important that this project be implemented *after* a JIT manufacturing system has been put into place, since a JIT system already requires highly accurate BOMs and a small number of suppliers. If that had been the case, there would have been no need to bring in a BOM engineer, and the number of expensive visits to suppliers could be reduced. If the presence of a JIT system had been assumed in this example, the cost of the project (assuming half the number of supplier visits) would have dropped to $122,005, which represents roughly a two-year payback. In short, efficient manufacturing systems must already be in place *before* payments based on production records are used.

In summary, these cost/benefit examples can be used to develop real-life examples, especially in terms of the line items used. The dollar amounts of the costs and benefits used in these examples were taken from the author's experience; however, cost and benefit projections should use amounts derived from the specific situations of the reader, not from these examples.

REPORTS

A typical accounts payable system has many reports related to checks that have been cut or payments due. Under the revised system described in this chapter, additional reports are listed that can be used as control points or to reduce the number of payments made. For example, a report describing changes in payment dates on long-term contracts allows the controller to anticipate payment changes and avoid making incorrect payments that require time to correct after the fact. Also, a form used to order manufacturing supplies from key suppliers does not directly relate to accounts payable (since it is a purchasing function) but leads to having fewer purchase orders to match against invoices, which reduces the work load of the accounts payable department. In short, a few extra reports and forms can improve the efficiency of the payables function.

One of the key control points in a revised accounts payable system is the purchase order. If a purchase order is stolen, it can be used to procure goods that will then be billed to the company. Though it is an after-the-fact control device, the following manual report is useful for spotting missing purchase order numbers; it lists the range of purchase order numbers used when printing a batch of prenumbered purchase order forms. Any missing purchase order numbers are immediately apparent, since there will be a gap in the numbering of successive purchase order batches. A typical purchase order numbering report is shown in Table 4.3. Note the gap in PO numbers between 3/9/98 and 3/10/98.

Of course, if the computer software contains the last purchase order number and prints the next number onto a blank purchase order, then this report is not needed. Instead, the controller should concentrate on tight computer security to keep anyone from printing a purchase order through the computer system; this requires no report.

TABLE 4.3   Prenumbered Purchase Orders Used

Date	Beginning No.	Ending No.
3/4/98	3400786	3400799
3/5/98	3400800	3400891
3/6/98	3400892	3400973
3/9/98	3400974	3401089
3/10/98	3401091	3401172
3/11/98	3401173	3401382
3/12/98	3401383	3401504

Another report that is useful in making payments on long-term contracts is the one shown in Table 4.4. It lists dates on which payments change. This may seem minor, but considerable time is expended whenever a payment is made in the incorrect amount, necessitating investigation, cancellation of the first check, and creation of a replacement check. An additional step worth considering (depending on the sophistication of the accounts payable system) is to avoid the report and load the change dates and amounts of the payment changes directly into the accounting software; then the payment changes will occur automatically. However, many accounting systems will not allow this information to be entered for future years.

The form shown in Figure 4.9 is useful for ordering supplies from a local office supplies company while avoiding the usual purchasing transaction cycle involving a purchase requisition and purchase order. On this form, the user simply fills in the quantity of each item needed and sends it to the office supply company. The description of each item as well as the part number used by the office supply company makes this an easy form for the office supply company to process. A similar form is shown in Figure 4.10 for manufacturing supplies. The key difference between these two forms is that there is usually more than one supplier of manufacturing supplies, necessitating a column on the manufacturing supplies form that lists the supplier to be contacted.

If a company pays suppliers based on the prices listed on the purchase order, then a key report in the process is a list of quantity and price disputes with suppliers. It can be used by the controller not only as a "memory jogger" of items to be resolved but also as a list of the potentially most troublesome suppliers. If this report consistently shows the same suppliers taking issue with the prices paid or the quantities used to calculate the total reimbursement, then it may be time to find a new supplier. The report shown in Table 4.5 lists the prices and quantities at issue in specific purchase orders, and is sorted by supplier.

TABLE 4.4   Payment Dates on Long-Term Contracts

Supplier Code	Supplier	Change Date	Payment Amount	Payment Frequency
ARCOAL	Argentine Coal Corp.	3/3/97	$100.32/ton	Monthly
ARCOAL	Argentine Coal Corp.	3/3/98	102.81/ton	Monthly
ARCOAL	Argentine Coal Corp.	3/3/99	105.22/ton	Monthly
TEGAS	Texas Gas & Mineral	12/31/97	.52/cubic foot	Weekly
TEGAS	Texas Gas & Mineral	12/31/98	.54/cubic foot	Weekly
TEGAS	Texas Gas & Mineral	12/31/99	.57/cubic foot	Weekly
TEGAS	Texas Gas & Mineral	12/31/00	.62/cubic foot	Weekly

SUPERFLOW TECHNOLOGIES
Office Supplies Order Form
Date _____

Deliver to _____

Description	Mfg. No.	Office Depot No.	Unit	Quantity
Binders				
Wilson Jones 8-$\frac{1}{2}$ × 11, $\frac{1}{2}$″ clear	70200	435-164	Each	_____
Diskettes				
DS, HD, 2MB 3M diskettes	12881	438-226	10/box	_____
Envelopes				
Clasp envelope, 10 × 13	COR56	822-940	100/box	_____
Mailing envelope, 10 × 13, white	CO925	423-731	100/box	_____
#10 business envelope, white	CO196	804-724	100/box	_____
Folders				
Globe-Weis letter file folder	321-1/3	475-814	100/box	_____
Globe-Weis legal file folder	322-1/3	475-806	100/box	_____
Pendaflex letter hanging folder	4152-1/5	449-421	25/box	_____
Pendaflex legal hanging folder	4153-1/5	433-425	25/box	_____
Mailers				
10-$\frac{1}{2}$ × 16 air bubble mailer	18518	806-729	12 pack	_____
Paper				
Copier paper, letter	DC11	477-299	5K/case	_____
Copier paper, legal	DC14	479-253	5K/case	_____
Copier paper, 11″ × 17″	3R3761	345-603	500/ream	_____
Phone message book	SC11542	366-732	2 pack	_____
Writing pad, letter size, white	7556	581-371	12 pack	_____
Quadrille pad, letter size	76581	302-356	12 pack	_____
Paper Clips				
#1 Regular	72380	808-881	1,000	_____
Jumbo	72580	808-907	1,000	_____
Small binder clips, $\frac{3}{4}$″	72020	808-857	12 pack	_____
Medium binder clips, 1-$\frac{1}{4}$″	72050	808-865	12 pack	_____

FIGURE 4.9 Office supplies order form.

A very useful payables control report is one that lists the amounts of finished goods in any time period, with payments to suppliers clearly indicated based on the amount of finished goods. A sample of such a report is shown in Table 4.6. The report lists the quantity of finished goods produced on a given day, the quantity of specific parts contained in the finished goods, and the price to be paid to suppliers based on the quantities of finished goods completed. This report can be used as detailed backup to payments in case suppliers complain about the amount

SUPERFLOW TECHNOLOGIES
Manufacturing Supplies Order Form
Date _____
Deliver to _____

Description	Supplier	Part No.	Unit	Quantity
Welding Supplies				
Weldwire, 308L, TIG Rod, $^1/_{16}$	High Plains	Weldwire	Feet	_____
Weldwire, 309, TIG Rod, $^1/_{16}$	High Plains	Weldwire	Feet	_____
Weldwire, 316L, TIG Rod, $^1/_{16}$	High Plains	Weldwire	Feet	_____
Weldwire, 308L, TIG Rod, $^3/_{32}$	High Plains	Weldwire	Feet	_____
Weldwire, 316, TIG Rod, $^3/_{32}$	High Plains	Weldwire	Feet	_____
Shop Machine Supplies				
Drill Bit, $^7/_{32}''$	PAI	90633	Each	_____
Drill Bit, $^{15}/_{64}''$	PAI	90634	Each	_____
Drill Bit, $^1/_4''$	PAI	90635	Each	_____
Drill Bit, $^{17}/_{64}''$	PAI	90636	Each	_____
Drill Bit, $^9/_{32}''$	PAI	90637	Each	_____
Drill Bit, $^5/_{16}''$	PAI	90639	Each	_____
Drill Bit, $^{21}/_{64}''$	PAI	90640	Each	_____
Drill Bit, $^{11}/_{32}''$	PAI	90641	Each	_____
Drill Bit, $^3/_8''$	PAI	91756	Each	_____
Hole Saw, Bi-Metal, $^3/_4''$	PAI	91068	Each	_____
Hole Saw, Bi-Metal $^7/_8''$	PAI	91070	Each	_____
Hole Saw, Bi-Metal 1″	PAI	91072	Each	_____
Hole Saw, Bi-Metal 1-$^1/_{16}''$	PAI	91073	Each	_____
Hole Saw, Bi-Metal 1-$^1/_8''$	PAI	91074	Each	_____
Hole Saw, Bi-Metal 1-$^1/_4''$	PAI	91076	Each	_____
Polishing Supplies				
Plug Deburr $^1/_2 \times$ 1-$^1/_2$	M.L. Foss	W221	Each	_____
Plug Deburr $^7/_8 \times$ 1-$^1/_2$	M.L. Foss	W188	Each	_____
Sand, Grit 80, Duralum	High Plains	ALOX0080	lbs	_____
Sand, Grit 120, Duralum	High Plains	ALOX0120	lbs	_____
Sand, Grit 240, Duralum	High Plains	ALOX0240	lbs	_____
Shop Supplies				
File, Flat Bastard 12″	PAI	90011	Each	_____
File, Flat Smooth 12″	PAI	90014	Each	_____
File, Multi-Kut 10″	PAI	90037	Each	_____
File, Multi-Kut 12″	PAI	91378	Each	_____

FIGURE 4.10   Manufacturing supplies order form.

TABLE 4.5 Payment Disputes with Suppliers

Supplier	PO No.	PO Line Item	PO Price	Supplier Price	Price Var.	Supplier Qty	Received Qty	Qty Var.
Davis	92456	17	$12.32	$12.56	+.24	1,000	998	−2
Davis	93405	42	72.31	75.01	+2.70	500	459	−41
Montford	92155	03	88.33	90.11	+1.78	250	249	−1
Montford	93602	01	90.14	90.22	+.08	1,500	1,504	+4
Montford	94111	02	67.54	75.75	+8.21	100	92	−8
Thimbles	91002	01	12.57	15.62	+3.05	100	92	−8
Thimbles	93067	05	1.45	3.00	+1.55	50	46	−4

TABLE 4.6 Amounts Due to Suppliers Based on Production Quantities

Finished Goods (F/G)	F/G Qty Made	Part Used to Make F/G	Part Qty	Extended Part Qty	Part Price	Extended Part Price
Wheel barrow	32	Wheel	1	32	$3.50	$ 112.00
		Hand grips	2	64	1.22	78.08
		Legs	2	64	5.58	357.12
		Bin	1	32	8.73	279.36
Shovel	1,114	Scoop	1	1,114	4.82	5,369.48
		Handle	1	1,114	2.73	3,041.22
Shears	2,458	Hand grips	2	4,916	1.22	5,997.52
		Shears	2	4,916	3.50	17,206.00
		Bolt assembly	1	2,458	.53	1,302.74

of payments made. This report can only be used if there is one supplier for every part; with more than one supplier, this report cannot itemize which supplier to pay.

If an automated system is used to process employee expense transactions, then the controller should know about instances when employees are not providing receipts for expenses that exceed the company-mandated minimum expense level. If properly designed, the automated expense processing system should allow the employee to enter a reason for any missing receipts. This information can then be rearranged into a report format for review. This report is very useful for spotting employees who are habitual offenders in terms of losing receipts; additional education of those employees, as well as one-on-one discussions, is usually sufficient to correct the problem. A typical missing receipt report is shown in Table 4.7.

If an automated expense reporting system is used, then a key performance measure related to it is the turn-around time required to get a payment back to the

TABLE 4.7   Missing Expense Receipts

Employee	Expense Report Date	Expense Report Line Item	Line Item Desc.	Amount	Explanation
Barney, George	1/17/97	04	Airline	$ 542.00	Lost wallet w/receipt
Barney, George	1/17/97	05	Car	148.00	Lost wallet w/receipt
Chumley, Fred	1/24/97	02	Airline	1,042.00	Briefcase stolen
Chumley, Fred	1/24/97	03	Hotel	493.00	Briefcase stolen
Davies, Alice	1/17/97	01	Meal	82.31	Lost the receipt
Davies, Alice	1/17/97	07	Airline	951.00	Lost the receipt
Eddings, Percy	1/24/97	13	Car	232.00	Lost the receipt

employee for expenses submitted to the company. The report shown in Table 4.8 lists the date on which an expense report was electronically submitted, the date the check was cut, and the time lapse between those two events. Also, since it is a major cause of delay in the payment of employees, the time required to obtain electronic approval from the supervisor is also listed (the start date for the supervisory approval is the same date as the electronic submission date, since the report is automatically copied into the electronic mail file of the supervisor as soon as the report is submitted).

In summary, the previous set of reports can be used by the controller for a variety of reasons—reducing the number of purchase orders requiring matching to invoices, spotting troublesome suppliers, listing employees who consistently cause problems on their expense reports, highlighting missing purchase order numbers, and so on. When used together, they allow the controller both to reduce the accounts payable department's work load and to highlight payables areas that are causing additional work to be performed.

TABLE 4.8   Expense Reports Transaction Times

Employee No.	Employee	Filing Date	Payment Date	Time Lapse (days)	Approval Period (days)
000045	Bailey, Brad	5/3/97	5/10/97	7	4
000105	Davis, Charlie	5/1/97	5/21/97	20	18
000271	Ermine, Nancy	5/9/97	6/4/97	26	23
000521	Fingal, Mandy	5/2/97	5/15/97	13	2
000947	Gretz, Nimrod	5/8/97	5/12/97	4	1
001004	Smith, Herbert	5/1/97	5/8/97	7	1
002467	Yantz, Alfred	5/4/97	5/22/97	18	7

TECHNOLOGY ISSUES

Many of the improvements advocated in this chapter involve the use of technology that has been available for a number of years. The primary improvement of this type is using a computer terminal in the receiving area to check receipts against the purchase order database and to record quantities received and shipping information on the spot. However, more recent technologies can make additional improvements to the accounts payable function.

Corporate Purchasing Card

An easy technological advance to implement is the corporate purchasing card. It is simply a credit card that can be restricted to maximum purchase amounts and purchases at specific suppliers. Also, the monthly statement can be itemized to identify items purchased, where the goods were purchased, and (sometimes) who made the purchase. These cards are of most use in reducing the work load of the purchasing and accounts payable departments, since the number of purchase orders and supplier payments decline when they are used.

Computerized Employee Reimbursement System

An innovation that usually requires extensive custom programming work is the computerized employee reimbursement system. This system copies the manual processing of employee expense reports, with a few variations; it eliminates paper-based expense reports and allows supervisors to approve the reports electronically (using a personal identification number) instead of with a signature. The system incorporates expense receipts by printing out a transmittal slip that lists all required receipts; the employee then attaches the expense receipts to the transmittal and mails it to the accounts payable department. The accounts payable department ensures that the correct receipts are attached and then approves payment. This automation is most useful for reducing the work load of the accounts payable clerk.

Optical Scanning of Payables Information

Another technological advance is optically scanning payables-related information into the computer system. This is very useful for record retention, since supplier invoices are frequently lost in transit when they are sent to supervisors for approval; if the document has previously been scanned into the computer, then it can be easily reprinted. Also, when suppliers call with questions about payments, the payables clerk can call up all related documents on screen while the supplier representative is on the phone, rather than going to the filing area, retrieving the documents, and then calling back the supplier (who may by then not be near a phone). In addition, if the scanned images are stored in a location that is linked to a network, then multiple users can review a scanned document at the same time. Finally, using a password on stored documents eliminates the need for locking file cabinets, since the password becomes the lock on the record in the database.

To begin a scanning transaction, a scanner with (preferably) an automatic document feeder is used. If two-sided documents must be converted, a scanner with a duplex scanning feature should be used. Scanned documents should initially be reviewed on a PC and then sent to storage on an optical disk. A large (perhaps 17-inch) monitor is useful for reviewing scanned documents. A computerized document storage system keeps the index key for each scanned image in a PC (sometimes called the index server) and the document stored in an image and text database (usually on a high-capacity optical disk "jukebox"). The index database is usually kept on a hard drive to improve access time. The user calls up the document from a workstation by entering the document's index key. Output can be to a monitor, fax, or printer. Corrections can also be made from a workstation. The system can also include a computer that performs validation checks on calculations as well as on the reasonableness of the data in the records. The typical scanning system off-loads erroneous or hard-to-read documents for manual review but is capable of processing most documents with minimal operator interference.

A key issue to resolve when setting up a paperless system is how to index the documents being scanned into the computer. The index is used by the database to find a document, so it is critical to use an index key that can be easily found. Nearly all databases allow the user to use multiple keys, so the user can find a document with several different key words. Also, a scanned document merely leaves a picture in the database rather than a set of words that can be scanned for retrieval purposes. However, when documents are scanned using optical character recognition (OCR) software, the picture is converted to words. This allows the user to conduct a search of all words in the database, which ensures that any document scanned into the database will be found. Locating a misfiled document quickly is an important issue, because one study has determined that it costs an average of $125 in labor costs to find a missing paper document, and recreating a lost document (about 7.5% of all paper documents get lost) costs an average of $350 in labor.

Once a scanned image has been converted by OCR software into a text file, the data can be compressed to take up much less storage space on a computer's hard drive than is the case with a scanned picture image. The resolution of scanned documents greatly affects the storage requirements of the system. A 100 dots-per-inch (DPI) scan of a single page requires far less storage space than a scan with a resolution of 300 DPI. Of course, images scanned at 100 DPI show more grain than documents scanned at higher DPI levels and are therefore more difficult to read. Even a picture image can be compressed; this is done by eliminating the "white space" on a page. The compression ratio is the compressed amount of storage divided into the uncompressed amount of storage. A ratio of 2:1 is typical.

The information presented here is only a summary of the document scanning and retrieval process. For more information, any of the following trade magazines should be consulted:

Enterprise Wide
Keyfile Corp.
22 Cotton Road
Nashua, NH 03063
(603) 883-3800

File Majic
Westbrook Technologies, Inc.
22 Pequot Park Road
Westbrook, CT 06498
(203) 399-7111

Imaging
12 West 21 Street
New York, NY 10010
(212) 691-8215

Imaging World
IW Publishing, Inc.
PO Box 1358
Westboro, MA 01580
(800) 344-6736

PaperClip
PaperClip Imaging Software
Continental Plaza 1
401 Hackensack Ave.
Hackensack, NJ 07601
(201) 487-3503

PCDOCS Open
PCDOCS Inc.
124 Marriott Drive
Tallahassee, FL 32301
(800) 933-3627

Recollect
Mindworks
735 North Pastoria
Sunnyvale, CA 94086
(408) 730-2100

SoftSolutions
SoftSolutions Technical Corp.
1555 North Technology Way
Building H
Orem, Utah 84057
(801) 226-6000

In short, document image storage and retrieval can be used to greatly simplify the document filing job of the accounts payable department. With quicker access to information of all kinds, the payables staff is able to research documents quickly, thereby increasing the speed of the average payables transaction.

Mobile Invoicing Systems

Finally, a recent innovation is to have the delivery person bring a hand-held device that can print an invoice for the customer on the spot or produce a floppy that contains the delivery information. In either case, the data should include part numbers and descriptions, quantities, and prices. This system eliminates the need for later rekeying of data by the clerical staff, since the delivery person and the receiving clerk inspect the parts together—if the delivery person's database is incorrect, then it can be changed and brought back to the supplier, who uses the altered information to change its database. This system does a good job of matching the information used by both the supplier and the company, so that there are no disputes over quantities delivered.

Bar-Coding of Shipments

One final technological device is not new—the bar-coding of shipments. If a shipment has the company's part number, unit of measure, purchase order, and quantity delivered already encoded into a bar code on the outside of the shipped carton, then a few seconds of scanning by the receiving staff will enter all this information into the company's database, thereby eliminating any possibility of errors caused by rekeying data incorrectly. It is a useful way to ensure the accurate transmission of data about a shipment from the supplier to the company.

In summary, technology allows the accounts payable department to reduce or eliminate supplier invoices, merge small purchases into single monthly credit card payments, eliminate employee expense statement analysis, enter receiving information with no keypunching errors, and have all payables-related information available from a computer monitor instead of from a filing cabinet. Even implementing a few of these systems would result in direct improvements in the speed of accounts payable transactions.

MEASURING THE SYSTEM

The traditional measurement most frequently used to determine the performance of an accounts payable department is the number of supplier invoices processed per person. Most of the measurements in this section relate to changes made before the supplier invoice ever arrives at the company (if it arrives at all). Thus, most of these measurements try to highlight the volume of transactions about to reach the payables department.

Number of Suppliers

One measurement is the number of suppliers. First, the company must add up all the suppliers who sent it invoices in the period just prior to the accounts payable and purchases changes being made. This is the baseline figure. Then, as the number of suppliers is reduced by using office and manufacturing supplies ordering forms, corporate purchasing cards, and supplier qualification techniques, the number of suppliers should be reviewed monthly and reported. To measure the most recent number of suppliers, simply measure the companies to whom checks were cut in the last month. It will not do to measure the number of companies for whom purchase orders were cut, because that assumes the company has converted to 100% purchase order usage, which may not be the case. The purchasing department should meet regularly to review the names on the current supplier list and find ways to incrementally shrink the list. As the list shrinks, the accounts payable department will find that it is consolidating more purchase payments into fewer checks, thereby reducing its work load.

Number of Payables Locations

Another measurement is the number of payables locations per billion dollars of revenue. A 1994 study found that an average company has 4.3 accounts payable systems per billion dollars of revenue. A world-class company (of any size) has one payables system. By reviewing the company's performance against this benchmark, management can continue to reduce the number of payables systems until the desirable level (one system) is reached.

Percentage of Payments with Purchase Orders

A very important measurement is the percentage of payments with related purchase orders. If the company is trying to move in the direction of paying suppliers from purchase orders instead of from supplier invoices, then this is the accounts payable measure to track. A small amount of programming will yield this measure—just summarize the total number of payments per month and divide that number into the number of payments that had a purchase order number in the payment record. The measure to strive for is 100% of all payments from a purchase order (with one exception noted in the next section). To improve this measurement, purchasing and accounting management should periodically review a listing of all payments that did not include a purchase order and incrementally implement additional controls that will ensure that purchase order numbers are issued the next time such transactions reoccur. Thus, this performance measure can be incrementally improved by continuing management attention.

Percentage of Purchase Orders below Minimum Value

Another key measurement related to the number of payments with related purchase orders is the percentage of purchase orders issued below the minimum threshold. A key point made earlier in this chapter is that the company should endeavor to purchase low-dollar items without purchase orders, in order to reduce the work load of the purchasing, receiving, and payables departments. The company must set up a dollar threshold below which no items should be issued a purchase order. The computer system can be programmed to report on every item below that level which was issued a purchase order number. The management team can meet periodically and review this list, implementing changes that will gradually reduce the number of these items by such means as including them in corporate card purchases or including them in standard supplies ordering forms.

Percentage of Total Payments from Production Records

If the company is moving to payments based on completed production information, then a key performance measure is the percentage of total payments made from production records. When backed up by a detailed report that lists the suppliers *not* being paid from production records, this information can be used by management to target additional suppliers who will be moved to payments from production records. Though it is nearly impossible to achieve 100% of all payments from production records (such items as rent and lease payments will never be paid that way), this performance measure can be constantly reviewed and incrementally improved.

Speed of Expense Report Turn-Around

When automated expense report processing is implemented, the time needed to manually complete the employee reimbursement transactions is minimal. Since employee-related costs are no longer of great concern, the best performance measure shifts into an entirely different area—the speed of expense report turn-around. This performance measure requires management to concentrate on getting payments to employees as quickly as possible, allowing them to pay their credit card companies on time, and thereby improving employee morale. The greatest problems found when using this performance measurement are the speed with which the accounting staff matches mailed-in receipts to the electronic report and (especially) the speed with which supervisors review the electronic expense reports and attach their approvals to the records. The measurement is calculated by using the filing date of the electronic expense report as the start date, and the date of check issuance as the termination date. Then the average processing period is derived from all expense reports in the period, with a detailed backup that provides information about specific expense reports that took an excessive amount of time to process.

Number of Accounts Payable Transactions

The last measurement to track is the number of accounts payable transactions per payables employee. This measurement is the last one to review, since it will vary depending on the improvements being tracked by the other measurements already highlighted in this chapter. A 1994 study found that a world-class company had 25,000 accounts payable transactions per year for each full-time accounts payable employee versus 12,000 transactions for an average company.[5] This information could be used as a benchmark to set against the company's amount per employee. Since there are many changes that "roll up" into this measurement, a company's management may use it as an overview of how the payables department is doing but cannot use it to track specific changes—it is too general a measurement.

Cost per Accounts Payable Transaction

Stated another way, the company can track the cost per accounts payable transaction. This is a more refined measure than the number of transactions per payables employee, since one can also focus on the related cost, which has a greater effect on the company's profits. The preceding study found that a world-class company could complete an accounts payable transaction for $1.80 versus a cost of $7.00 for a typical Fortune 100 company. Again, this is an overview measure, since it does not lead management to make specific improvements in operations—it is too general a measurement.

In summary, most of the performance measures noted in this section are related to incremental and continuing improvements in the payables process. For example, the percentage of purchase orders issued below the minimum threshold

---

[5] "Accounting Practices Benchmarking Study Spots Mistakes Companies Make," *Journal of Accountancy,* March 1995, p. 24.

requires management to continue to reduce the number of low-cost purchase orders. Another key measurement, the expense report turn-around time, leads management in the direction of constantly shrinking the time required to get payments back to employees. By constantly reducing these transaction times, the controller increases the capacity of the accounting department to perform other tasks.

IMPLEMENTATION CONSIDERATIONS

A number of problems relating to change management may arise when a revised accounts payable system is installed (see Chapter 11). The most common implementation issues related to a faster accounts payable system are the following.

Long-Term Contracts, Service Agreements, and Emergency Services

When purchase orders are required for all purchases, the controller must be prepared for situations involving long-term contracts, service agreements, and emergency services. Blanket purchase orders should be issued for long-term contracts and service agreements. Though purchase orders can rarely be issued when emergency services are required, they can be issued after the fact to bring emergency service providers into compliance with company policy. Since one of the reasons for having a purchase order is to allow the computer system to automatically pay from information in the purchase order database, it is acceptable to create a purchase order even after the emergency service has been provided. Of course, late purchase orders violate the principle of having advance management approval of purchases, but emergency services are difficult to avoid.

Control over Purchase Order Forms

Controlling purchase order forms is difficult if the staff is used to storing them in a convenient spot. Requiring them to lock up excess forms and to track purchase order numbers will be regarded as a "pain." The best way to get around this problem is to have the internal auditing group review the situation frequently and report transgressions to senior management.

Reducing the Number of Suppliers

Centralizing the purchase of office and manufacturing supplies with a very small number of suppliers is welcomed by the staff. However, the controller will find that some employees have favorite products that can only be purchased from other suppliers and that therefore entail paying for extra invoices and matching them with extra purchase orders. A review of invoices each month will reveal these small-dollar items; such items should be referred back to the purchasing department, which can usually find comparable items through the standard suppliers.

Excessive Corporate Card Use

When corporate purchasing cards are used, there may be employees who abuse the system by purchasing too many supplies and retaining some supplies for personal use. If the internal audit staff periodically reviews purchases made by employees, abusers will have their cards taken away, thereby solving the problem.

Paying Based on Purchase Orders

The biggest implementation issue is paying based on purchase orders instead of supplier invoices. Internally, the controller must be involved with controls built into the accounting and purchasing software to ensure that suppliers are paid for the amounts received. Specifically, the software must allow the receiving clerk to call up the supplier purchase order on the screen and enter the quantity received into the computer system. Externally, suppliers must be trained to prominently mark the purchase order number on all goods shipped to the company or mark it on any attached shipping documents, so that the receiving clerk can enter it into the computer system to verify that a purchase order was issued.

Also, few suppliers are used to having payments made to them based on purchase orders. Most suppliers prefer to receive a check with their own invoice numbers listed next to the payment, so that they can offset the appropriate line items in their accounts receivable records. Because suppliers' accounts receivable procedures and computer software are focused on receipts based on invoices, they will resist payments based on purchase orders. The best way to solve this problem is to send out written notification of the change with several months' warning. Key suppliers may require personal visits to explain the situation; the worst way to deal with this implementation issue is to start paying based on purchase orders with no warning at all. Also, a contact name should be given to all suppliers whom they can call about pricing or quantity disputes under the new system. If suppliers request extra time so that they can program their receivables systems to process cash based on a customer purchase order number, then a moderate time period should be granted to them.

Shipping Charges

Paying suppliers from purchase orders is easier when there are no shipping charges to pay, since these costs are not readily discernible without a supplier invoice. To avoid the problem, suppliers can be contacted and asked to build the shipping price directly into the prices of the materials (which *are* listed on the purchase order). Alternatively, suppliers can be asked to have shippers charge the company directly. In either case, advance preparation is the best way to deal with the problem.

Sales Taxes

Paying taxes is a difficult issue that arises when paying based on purchase orders. Once again, the best approach is advance preparation. The company can either procure software that lists all the tax rates of all governments throughout the country, or (and more easily) have the supplier provide local tax rates, which are then listed on the purchase order record in the database. With the information in the database, the accounts payable system can easily calculate the amount to be paid. This usually requires custom programming and should therefore be planned well in advance.

Paying Based on Production Records

Seemingly the simplest payables system is paying suppliers based on the quantity of finished goods the company manufactures; in reality, it is the most complicated to implement. One problem is that the bills of materials must be absolutely accurate before the new payables system is implemented or else payments to suppliers will be far from accurate. To solve this, all bills of materials must be reviewed by a team of engineering and production personnel to verify their accuracy. Then the bills must be audited regularly by the internal auditing department to ensure that accuracy levels are being maintained. Finally, procedures must be put in place that allow changes to be made to the bills in an orderly manner.

Another problem with paying from production records is that suppliers are unhappy with this method—they (rightly) feel that it is too easy for things to go wrong and for payments not to be made. Consequently, it is best to implement the system with just one supplier and to work out the "bugs" in the process before expanding the process to other suppliers.

Yet another issue with paying from production records is that not all received items are incorporated into the company's products; thus, there will always be items that must be received, entered into a receiving log, and paid through a more conventional method. Consequently, the controller should budget for a staff presence in the receiving area even after the new payables system is installed.

Automatic Expense Report Processing

Automatically processing expense reports through the company's computer system involves two implementation issues—programming and user training. Programming is perhaps the more difficult issue, because this application is a custom programming project—it is not yet commonly seen with standardized accounting packages. Consequently, the controller must budget a very considerable time period for the requirements definition, programming, and debugging phases of the installation. Also, to ensure that the employees use the software correctly, all employees who will be submitting electronic expense reports must receive training on the system; accordingly, the controller should budget both employee time and training fees to complete the training process.

In summary, there are implementation issues associated with virtually every payables process improvement presented in this chapter. By using controls over access to forms, employee training, supplier training, emphasizing the accuracy of bills of materials, and budgeting sufficient time to create customized software programs, the controller can implement these changes.

SUMMARY

This chapter reviewed transactions required to process a payable in a typical company and then found a number of ways to reduce the time required: a single approval point (instead of three), payment from a single document (instead of three), and requiring no reconciliation of conflicting documents. This improvement was achieved with computer technology that has been available for several years and is a classic example of how to use modern computer systems to avoid much of the transaction-processing time needed in a traditional accounting process.

REFERENCES

Anderson, Ernest L., Jr. "The Accounts Payable Link." *Purchasing,* April 1993, pp. 31–32.

Coburn, Steve, Hugh Grove, and Cynthia Fukami. "Benchmarking with ABCM." *Management Accounting,* January 1995, pp. 56–60.

Elgin, Peggie R. "T & E Payment Options Trim Reconciliation Efforts." *Corporate Cashflow,* October 1993, pp. 6–11.

Hackett Group. *AICPA/THG Benchmark Study: Results Update and Analysis.* 1994.

Hunton, James E. "Setting Up a Paperless Office." *Journal of Accountancy,* November 1994, pp. 77–85.

Image Processing and Optical Character Recognition. New York: AICPA, 1993.

Queree, Ann. "It's a Buyer's Market." *Corporate Finance,* March 1994, pp. 14–18.

Singhvi, Virendra. "Reengineer the Payables Process." *Management Accounting,* March 1995, pp. 45–49

Tully, Shawn. "Purchasing's New Muscle." *Fortune,* February 20, 1995, pp. 75–83.

Weiss, Mitchell Jay. "The Paperless Office." *Journal of Accountancy,* November 1994, pp. 73–76.

Whittington, O. Ray, and Kurt Pany. *Principles of Auditing.* Chicago: Irwin, 1995.

Willson, James D., Janice M. Roehl-Anderson, and Steven Bragg. *Controllership.* 5th ed. New York: Wiley, 1995.

5

COST ACCOUNTING

The cost accountant typically cannot begin work on period-end closing tasks until many other tasks have been completed, since information from other accounting functions is used as input into cost accounting reports. Since there is usually a deadline for completing the financial statements, the cost accountant is "between a rock and a hard place"—the costing work cannot start until other work is finished, and the preceding work may not be completed on time. Consequently, the cost accountant works under extreme pressure to provide explanations for what are frequently the largest variances on the income statement—those related to the cost of goods sold. One solution is to extend the financial statement due date until the variance analysis is complete. This chapter discusses other techniques to assist the cost accountant without extending the financial statement due date.

In addition, this chapter discusses more effective measures than the period-end variances for the cost accountant to track. Examples of these are inventory, labor routing, and bill of materials accuracy. These serve the double purpose of providing more effective management information and of moving the cost accountant's tasks away from the period-end closing into other portions of the accounting period.

CURRENT SYSTEM

During a typical accounting close, the cost accountant is forced to wait while other accounting functions are completed. Then a specific set of variances must be calculated, their causes investigated, and additional investigations initiated if the first round of explanations is not adequate. This investigation is usually cursory, since the cost accountant is under considerable time pressure to deliver a reasonably plausible variance analysis to the controller, and the production personnel being questioned know this—if they give reasonable answers (e.g., excessive scrap due to some other department's problems, or extra overtime to achieve the month's shipping goal), then the cost accountant will go away satisfied. After the month's results have been published, the cost accountant will make some attempt to verify the analysis, but the next period end will be arriving in a few days, and then the cost accountant will abandon the last set of variances and concentrate on the variances for the next month. This cycle of activities does not allow the cost accountant to conduct in-depth analyses and does not contribute to fixing the underlying problems causing the variances. This section reviews this process in some detail.

To begin, Figure 5.1 shows the accounting functions that must be closed before the cost accountant can begin to analyze variances. The accounts payable function must close, because invoices for additional inventory items must be added to the month-end inventory balance, and these bills can arrive several days after the end of the month. If the cost accountant were to begin work before this area closed, the book ending inventory balance could be too low. Next, the receivables function must close, because that closing process includes a review of the shipping log to ensure that all items shipped have been logged out of the system (which reduces the inventory balance). If the cost accountant were to begin work before this area closed, the book ending inventory balance would be too high, causing several variances to occur. Finally, the inventory function must close for the month. For companies that have either periodic physical inventory counts or perpetual inventories, this involves a reconciliation of the physical amounts to the book balances, involving a further reconciliation of costs as well

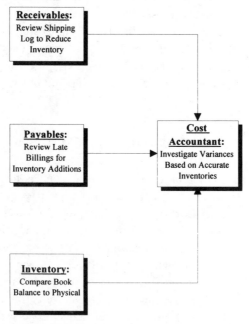

FIGURE 5.1   **Accounting closings preceding the work of the cost accountant.**

as quantities. Without this information, the cost accountant will have great difficulty in determining the causes of variances.

In those companies who close their inventory areas each month without a cross-check against physical balances, the cost accountant can begin work before the inventory area closes, because the information provided by the inventory function will be of no use to the cost accountant—any materials variance explanations are guesses without concrete inventory information, so the cost accountant will be just as uninformed after the closing of the inventory function as before it closed. The example in Figure 5.2 illustrates this problem.

The cost accountant completes other work besides materials variance analysis.

Analyze Labor Variances

The major labor variances are for price and efficiency. The cost accountant calculates them as part of the period closing process. The information is calculated during the closing period and is given to production management, which is expected to shift the production staff to keep the cost of labor in line with the budget as well as to use the efficiency analysis to fix problems that created excess labor usage in the previous period. In reality, the labor pool is relatively fixed, and the efficiency variance arrives too late to be of use to production management.

Analyze Overhead Variances

The major overhead variances are related to too much overhead being incurred (the spending variance), too much overhead being applied because too many base hours are being worked (the efficiency variance), and too much overhead being applied because additional units of product are being manufacturing (the volume variance). This information is calculated during the closing period and is given to plant management and production management. Only the spending variance is of any use to production management, since it indicates too much money spent on specific overhead line items. The other overhead variances are an interesting academic exercise for the controller to review, but production management can do little to alter the remaining variances except change the level of production (which may be driven by a number of factors besides overhead cost control).

Provide Closed Job Reviews

One of the more useful roles of the cost accountant is the analysis of revenues and costs for completed jobs or production runs. This information can be calculated outside of the usual closing period and is given to production management. It contains the amount of costs incurred, and may do so in great detail, depending

The controller of the XYZ Wholesale Products Company has nearly completed its January close and is calculating its ending inventory balance. Its beginning inventory is accurate at $914,000, because it was verified by the year-end physical count. The end-of-period inventory of $1,038,000 was derived by adding all purchases ($263,000) during the month to the beginning balance and then subtracting the items shipped out ($139,000), according to the shipping log. The cost of the items shipped out was determined by using the bills of materials for those items. The calculation is illustrated as follows:

Beginning Inventory	$ 914,000
+ Purchases	263,000
− Shipments	139,000
Ending Inventory	$1,038,000

The President of XYZ mentions to the controller that the upcoming year-end inventory had better not require the large (and disappointing) downward adjustment needed after last year's physical inventory. The controller responds that it is impossible to provide an ongoing materials variance analysis without a physical ending inventory each month. To prove it, he assigns the cost accountant, Mr. Forlorn, the task of providing a set of materials price and usage variances for January and documentation of how each variance was calculated without the benefit of a fully costed ending inventory. After a week of work, Mr. Forlorn, with an even longer face than usual, delivered the following report:

The materials usage variance was $1,500. Mr. Forlorn had rooted through the scrap bin, dug out a number of discarded items, and individually costed them with the help of the purchasing staff. This had taken 20 hours. He recommended setting up a scrap reporting form on which, every time an item was thrown out, the quantity, part number, and reason for scrapping were to be entered. Also, a large number of items had been returned to suppliers; another form was needed to track these returns.

The materials price variance was $7,000. Mr. Forlorn had reviewed all large-dollar purchases received during the month and manually checked these costs against the preset standards. This had taken 20 hours. He noted that no one was checking actual prices against standards and that a database should be set up to report on the variance.

When the president and controller reviewed this information, they decided to maintain a database of purchase costs and to apply these costs to an ongoing set of perpetual inventory records. They determined that the effort required to do this equaled the effort required to research the materials usage and price variance reports.

FIGURE 5.2   Example: Effect of inaccurate inventories on variance analysis.

on the amount of work allowed for the review. The information is used to plan for the costs and prices of similar projects or production runs in the future.

Consolidate Subsidiaries

An additional task for a corporate-level cost accountant is to combine the results of the cost accounting systems of all subsidiaries. One study shows that the average company has 3.9 cost accounting systems per billion dollars of revenue, so the consolidation process alone can be a full-time job. The information is given to senior corporate management for review. From a day-to-day management perspective, the information is of no use, since the information is retained at the corporate level.

In summary, the majority of cost accounting tasks are compressed into a small time frame during the closing process, as shown in Figure 5.3. This results in hastily formulated variance reports that are so highly summarized that the information is of little use to production management, which needs detailed information to solve variance problems. Also, the information is provided much too late for production management to solve issues; detailed variance information is required every day, not once a month. The cost accountant must reduce the number of tasks performed during the closing cycle and issue variance reports daily, so that fresh information is provided to the production staff.

REVISED SYSTEM

A revised cost accounting system should tackle the issue of speeding up the accounting process by addressing not only the tasks occurring during the closing process but also the relevance of the information transmitted to the operational staff. This allows the company to become more efficient (by speeding up the process) and more effective (by providing information that can be used to improve operations).

The search for more relevant information has the secondary result of shrinking the number of cost accounting tasks conducted during the accounting close. This section first suggests cost accounting tasks that create better information and then examines how these affect the speed of accounting transactions.

Review Inventory Accuracy

One of the primary flaws in most cost accounting systems is the lack of a true perpetual inventory system. When the beginning or ending balances are incorrect, the cost of goods sold is incorrect. To avoid this problem, the cost accountant

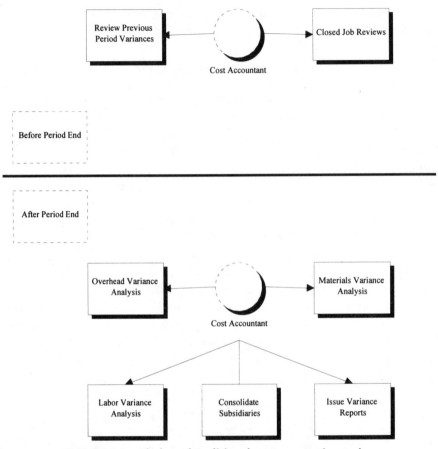

FIGURE 5.3   Timing of traditional cost accounting tasks.

should review a large proportion of the inventory on a continuing basis and verify that the inventory has an accuracy level of at least 95%. This will help to reduce the materials quantity variance.

Review Bill of Materials Accuracy

Perhaps the greatest problem of all for the cost accountant is not having bills of materials that are at least 98% accurate. An inaccurate bill can be used to purchase inaccurate quantities of materials or the wrong materials entirely. Furthermore, the cost accountant uses bills of materials to cost products. Thus, if the bills are wrong, the costs are wrong. To avoid this problem, the cost accountant should be responsible for frequently reviewing bills of materials with the engineering and production staffs to verify that they are accurate.

Review Labor Routing Accuracy

Management cannot accurately plan for future labor staffing if the labor routings associated with products are incorrect. This leads to labor price and efficiency variances, since incorrect staffing leads to having either over- or under-qualified staff (with associated pay rate variances) conducting either too little work (leading to negative efficiency variances) or too much work (costing the company in overtime expenses). To avoid this problem, the cost accountant should be responsible for frequently reviewing labor routings with the engineering and production staffs to verify that they are accurate.

Review Targeted Materials Costs versus Actual Costs

If suppliers do not charge for materials at prices that are agreed upon in contracts, materials prices can spiral out of control. To avoid this problem, the cost accountant should compare actual materials prices to those noted in supplier agreements and report any variances to senior purchasing management. The cost accountant should also be involved in the initial setting of purchased materials cost goals.

Review Inventory Obsolescence

Another cause of materials variances is the all-too-common write-off of large quantities of inventory during the year-end audit. This painful episode can be avoided by frequent reviews of the inventory by the materials review board (MRB), of which the cost accountant should be a member. The MRB usually has members from the purchasing, engineering, production, and accounting departments.

Review the Average Labor Rate

Over the long term, the average labor rate may go up or down, which may point to an unplanned change in the mix of experienced (and costly) staff to inexperienced (and less costly) staff. To avoid this problem, the cost accountant should be in charge of tracking the average labor rate.

Review Overtime Percentage

The amount of overtime worked may point to changes in the number of production employees or to the need to change the amount of production volume. The cost accountant should be responsible for tracking the overtime percentage.

Review Individual Overhead Costs

A limited number of overhead costs are highly variable and can cause the amount of uncapitalized overhead to jump unexpectedly. The cost accountant should delve into the reasons for changes in these variable overhead costs during each closing period.

Review the Amount of Overhead Capitalized

Substantial changes in either production volumes or actual overhead costs may require a change in the overhead application rate, which will avoid long-term overhead volume or spending variances. To spot these changes as early as possible, the cost accountant should track the amount of overhead capitalized in each reporting period.

The effect of these new tasks on the cost accounting job are shown in Figure 5.4, where many tasks have been moved out of the closing period and into the preceding portion of the month. The new set of cost accounting tasks will help to bring more relevant information to corporate management. But how will these tasks improve the *speed* of accounting transactions? Most of the new tasks not only can be but should be conducted all through the month rather than during the period-end close. This means that the time constraints imposed by the period-end closing process are removed from the cost accountant (except for reviewing individual overhead costs and overhead capitalization, which are best handled during the accounting close). As a result, the period-end close is conducted more quickly, since there are fewer cost accounting tasks to be performed at that time.

Does this mean that all the traditional period-end variance measures should be thrown out? No. These measures have been used by cost accountants for many years, and we should not be too quick to ignore the weight of so much tradition. A good reason for continuing to use the various overhead, materials, and labor variances is that they provide an excellent overview of where costing problems are occurring. If a new cost accountant has to design a costing system and wants to find out where the largest cost variances may be found, a quick review of the traditional cost accounting variances may reveal that the area of analysis should be, for example, labor efficiency, which would then lead the cost accountant to a review of labor routings. However, once the detailed costing systems are in place, it is not always necessary to calculate all these variances during the period-end closing process—they can, depending on the circumstances, be pushed out past the closing process, when the cost accountant can dissect the variances at leisure and investigate specific costing problems.

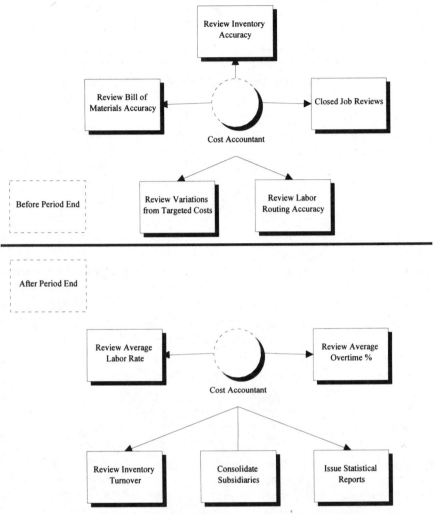

FIGURE 5.4 Timing of revised cost accounting tasks.

What are the circumstances under which the cost accountant will feel safe in putting off or ignoring an analysis of the overhead, labor, and materials variances? The following circumstances indicate a minimal need for the variances.

Overhead

Review the accounts that make up the overhead pool, and split the costs into fixed and variable elements. Usually, the bulk of the overhead costs will be such fixed costs as salaries, benefits, rent, depreciation, and utilities. Frequently, the only variable cost of any significance will be manufacturing supplies. If so, then it is easier to refer directly to the major variable costs as listed in the general ledger rather than compute the overhead spending variance.

If production volumes are relatively stable, the overhead volume variance is unlikely to fluctuate much from period to period. Instead, the cost accountant should track the difference between the actual overhead cost each period and the amount capitalized through the production process; any continuing difference between the two should be eliminated by a change in the overhead rate. Alternatively, if production volumes vary a lot from period to period, then the overhead volume variance should be reported.

Labor

Many companies have a stable work force that does not vary much from period to period. That being the case, a company's labor costs should be considered more fixed than variable. However, labor efficiency variances presume that the staff is laid off the moment there is no work left to complete. In reality, the staff is on site and being paid even if the work load drops substantially. Consequently, the labor efficiency variance is not entirely realistic when labor is a fixed cost. The cost accountant would be of more assistance to management by tracking the one element of the labor expense that *is* variable (sometimes in large amounts) from period to period—the amount of overtime paid as a percentage of the total labor cost. Overtime information can be used by management to make long-term decisions regarding the number of production workers and short-term decisions regarding the volume of work moving through the production facility.

In addition, the cost accountant should review the average labor rate to see if long-term swings in personnel are occurring. This information can be used by management in determining whether the production force is too highly paid for the skill level needed to perform the production work.

Materials

If a company has highly accurate bills of materials (at least 98% accurate) and an accurate inventory tracking system (at least 95% accurate), then materials quantity variances are easy to spot without the need for a high-level materials quantity variance analysis. However, if this kind of information is not available (which is frequently the case), then the variance should be reported—though the cost accountant will have a difficult time providing a detailed explanation for the variance because of lack of adequate information.

The final variance, the materials price variance, is the one most deserving of publication. A well-managed company will find that, depending on the industry, nearly all its costs are fixed in the short term except for materials prices. These can be tied to long-term supplier contracts that guarantee specific prices based on sole sourcing for long periods. If the cost accountant tracks variances from the prices noted in supplier contracts, then cost savings can be immediate. However, if a company is a custom manufacturer, long-term materials contracts are rare; as a result, materials costs cannot be locked in, since a wide range of suppliers are generally used to supply a broad range of parts that cannot be predicted in advance with any accuracy.

In short, in most cases, the cost accountant should review only the most variable of the overhead costs, the average overtime rate, and the materials price variance, and have reasonable control over short-term costs. Over the longer term, a comparison of the actual overhead cost to the amount capitalized, as well as a review of the average labor rate, will keep the cost accountant in touch with corporate costs. However, these reduced levels of control are sufficient only if other day-to-day measures are carefully controlled, such as the accuracy of inventory, labor routings, and bills of materials.

A key development in manufacturing systems that is of concern to the cost accountant is just-in-time (JIT) manufacturing. This system only produces enough product to meet customer needs and emphasizes the reduction of cycle time, since employees can then produce quickly while still maintaining minimal levels of work-in-process. Since the collection of variance information during the production process increases cycle time, variance analysis is discouraged in a JIT system. Without variance analysis, how does a JIT facility discover problems in the process? The key to JIT manufacturing problem resolution is the near-absence of work-in-process (WIP). As shown in Figure 5.5, a machine operator in a traditional assembly line process can continue to manufacture parts of the wrong size without anyone discovering the problem, since the parts are loaded into a WIP queue and may not reach the next workstation for several days. By the time the problem is discovered, a large number of parts will have been produced and will then have to be scrapped or reworked. However, as shown in Figure 5.6, a JIT system forces the same machine operator to take the part to the next workstation and process it further, at which time the defect will be discovered. Thus, under a cellular manufacturing system, only one part will be defective before the problem is fixed. If traditional variance reporting had been used in this environment, the cost accountant would have taken too long to collect the information, analyze it, and report the information back to management; by then, with no WIP to hide the problem, management would have already found and fixed the error.

The JIT system has removed the need for most variance reports by the cost accountant, so the cost accountant's focus turns to becoming involved in setting a target cost during the design stage and assigning targeted subsidiary part costs to suppliers. The cost accountant then reports on variances caused by suppliers' not meeting their cost targets.

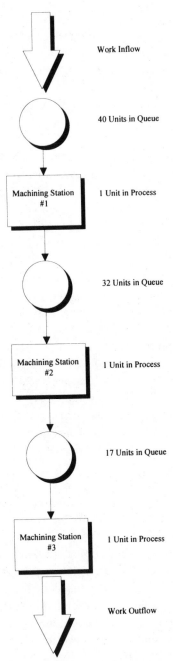

FIGURE 5.5   Volumes of work-in-process in an assembly line environment.

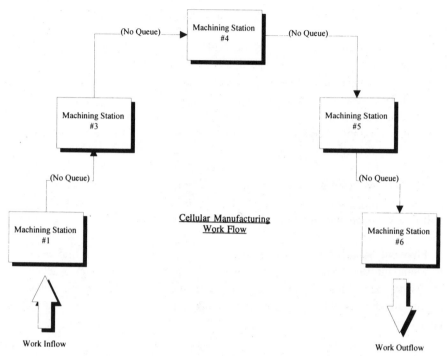

FIGURE 5.6   Volumes of work-in-process in a JIT environment.

In summary, the relevance of the information provided by the cost accountant is greatly improved by switching away from monthly variance reporting and into a review of the accuracy of such manufacturing data as the accuracy of inventory, bills of materials, and labor routings. Also, the cost accountant should advocate a switch to a JIT manufacturing system, so that the bulk of the variance analysis can be performed by the production crew instead of the cost accountant.

CONTROL ISSUES

There are many control issues related to cost accounting. For example, if the cost accountant eliminated or reduced the level of examination of certain variances, a knowledgeable person could obtain company assets, knowing that the variance that flagged the theft was no longer being reviewed. This section contains some examples of such situations, along with possible solutions to the control problems. The solutions do not involve period-end analysis; they can be conducted at any time, thereby allowing the cost accountant to complete period-end tasks quickly and move on to other work.

Labor Efficiency Variance—Embezzlement

This variance shows the efficiency of the labor force versus the standard labor time expected to be used in production. The cost accountant conducts a minimal review of this variance, because the production staff is considered a fixed cost and will be paid even if they are not efficient. A production supervisor realizes that the variance is not being reviewed and continues to clock in an employee who now works for another company. The paycheck is sent to the ex-employee by direct deposit, and the ex-employee splits the check with the production supervisor.

Physical Matching—Audit.   The controller can ask the internal audit department to physically match personnel to personnel records. This can be a difficult review, since personnel may be off-site.

Physical Matching—Continuing.   The controller can stop direct deposits and hand out the checks directly to the staff, marking down recipients on a master list to ensure that no one receives more than one check. Employees may not appreciate this method, however, since it takes away the convenience of direct deposits.

Labor Efficiency Variance—Overstaffing

Once again, the cost accountant conducts a minimal review of this variance. Realizing the lack of oversight, the production supervisor keeps more staff on hand than is required by the production schedule, so that the production schedule can still be met if sudden production increases are required. This costs the company money, since the extra personnel must be maintained during slow periods, but it makes the production supervisor look like a top-notch manager when product must be shipped on time.

Bonuses.   The production supervisor can receive a bonus incentive tied to the level of production staffing, so that the supervisor will lose money if overstaffing occurs.

High-Level Review.   Though the cost accountant may not conduct a detailed review of the labor efficiency variance, this does not mean that a review should not be conducted. The cost accountant can simply calculate labor as a percentage of material costs (or some other base that is closely tied to labor) and investigate large variances. However, any fundamental change in the business that causes large shifts in the amount of production labor will render this technique ineffective.

Labor Rate Variance—Mix Change

The cost accountant conducts a minimal review of the labor cost per person, since this is considered a fixed cost. The production supervisor realizes this, and replaces all low-paid, inexperienced staff with highly paid, skilled workers; this makes the production crew (and hence the supervisor) look very efficient, but is expensive for the company.

Bonuses.   The production supervisor can receive a bonus based on the average wage received by the production staff. This should be used with care, since the incentive may push the production supervisor in the opposite direction, to hire a "green" crew.

High-Level Review.   The cost accountant can very quickly calculate the average payroll cost per person each month and investigate any variances. However, this method can be rendered ineffective for a short period when large layoffs or bulk retirements (caused by early retirement programs) cause the baseline to change.

Age Review.   Assuming that older employees earn more pay, the cost accountant can calculate average employee age from payroll records. Any sudden jump in the average age can indicate a shift to a more experienced staff.

Materials Usage Variance

This variance shows the difference between expected and actual materials usage. The company has restructured its manufacturing processes so that cellular manufacturing has replaced independent workstations. As a result, work-in-process levels have dropped significantly, allowing management to focus on scrap problems. After much work, scrap has been reduced to the point where the cost accountant has decided not to conduct detailed reviews of wasted materials. Workers in several cells become less careful in their work, resulting in increased levels of scrap. Realizing that the materials variance is no longer being reviewed, they throw away the scrapped materials and do not report the loss.

Bonuses.   Production employee bonuses that are tied to scrap levels will help to keep scrap down without any variance analysis by the cost accountant.

High-Level Review.   If the company has accurate bills of materials, the cost accountant can create a report that multiplies the number of units shipped by the parts listed in each product's bill of materials, yielding a costed list of parts that should have been used during the period. When summarized, the total cost from this report can be compared to the actual materials cost for the period as a reasonableness check. If there is a notable variance, the cost accountant can then compare the usage quantities on the report to the actual quantities used for each part. This allows the cost accountant to highlight specific part usage, which can frequently be tied to the actions of specific production processes that use those parts. As long as inventory and bill of materials records are accurate, this information can be analyzed at any time of the month, not only during the closing process.

Materials Price Variance

This variance shows the difference between the actual and budgeted materials cost. This is the primary focus of the cost accountant, since it is the most variable item within the cost of goods sold. This area should not be subject to reduced review; on the contrary, the cost accountant may move to an even earlier step in the process and become involved in the cost of parts as they are engineered into a new product design. This action keeps costs out of the product before the first design is ever sent to the procurement department for sourcing with suppliers.

Overhead Spending Variance

This variance shows the difference in actual overhead costs from the budgeted amount. The overhead spending variance is caused by costs being incurred in excess of those budgeted for the period. When overhead costs are applied to a product at a predetermined rate when the actual overhead rate should be higher, a spending variance occurs. This is not a variance that a cost accountant can overlook, because changes in overhead costs must be constantly monitored and promptly reported to management. Since overhead continues to take a larger proportion of corporate costs, the cost accountant should be vigilant in watching over these costs. This is too important an area to ignore. However, when attempting to close the last period's books more quickly, this item can be safely reviewed after the period has been closed.

The cost accountant who tries to shorten the closing process, however, will try to move the cost approval process for overhead to a point earlier in the transaction cycle. This will reduce the need for a spending variance analysis, since all costs will have been reviewed well before any money was spent on the overhead items. For example, the cost accountant can recommend that all overhead items be listed on a purchase requisition that must be signed by several managers; this requires a review of costs before the costs are incurred instead of after the check has been mailed to the supplier. As a further enhancement of the process, each manager who signs a requisition can be given a report that shows spending to date in various cost categories, so that new requisitions may be rejected if the year-to-date budget will be exceeded if the expense if approved.

Overhead Efficiency Variance

This variance shows the difference between the standard overhead that would have been applied if labor had exactly matched the standard, and the overhead applied based on the actual amount of labor. The cost accountant has chosen not to track this variance. The amount of labor starts to rise, which leads to a similar increase in the amount of overhead charged to products.

Activity-Based Costing (ABC). The overhead efficiency variance is founded upon the belief that all overhead must be allocated using labor. An activity-based costing system uses multiple allocation bases, so that costs may be much more precisely assigned to projects. Use of ABC nullifies the need to track the overhead efficiency variance.

Overhead Volume Variance

This variance shows the difference between the standard overhead that would have been applied if the production volume had matched the budgeted volume, and the overhead applied based on the actual amount of production units. The cost accountant elects to ignore the overhead volume variance. Knowing this, the production supervisor decides to manufacture too much product, so that extra overhead will be absorbed, thereby creating a profit by capitalizing overhead into inventory.

Bonuses. As usual, employee behavior can be modified by offering bonuses if certain criteria are met. In this case, the production supervisor will receive a bonus if production exactly matches demand (or even shrinks the amount of on-hand inventory).

Review Inventory Levels. The cost accountant can review WIP and finished goods inventory levels to see if they are increasing. Higher WIP and finished goods inventory levels indicate that overhead is being capitalized into inventory. This review can be done at any time of the month, not only during the closing cycle.

Create a Volume Variance Table. A volume variance table is an easy way to quickly determine the expected gain or loss when overhead and production are

TABLE 5.1   Volume Variance

Overhead Rate	Production Level				
	$200,000	$225,000	$250,000	$275,000	$300,000
100%	$ 0	$ 10,000	$ 20,000	$ 30,000	$40,000
90	−10,000	0	10,000	20,000	30,000
80	−20,000	−10,000	0	10,000	20,000
70	−30,000	−20,000	−10,000	0	10,000
60	−40,000	−30,000	−20,000	−10,000	0

at certain levels, all other variables being equal. An example of a volume variance table is Table 5.1. The cost accountant can use this information to obtain a high-level estimate of the overhead volume variance without being burdened by an excessive amount of variance analysis. The table shows a zero variance in its upper left-hand corner; this position uses the budgeted overhead and production volumes that were set at the beginning of the budgeting period. As production volume rises above the budgeted level, the cost accountant can predict the amount of volume variance and recommend a new overhead rate to the controller that will eliminate the volume variance. For example, if the cost accountant sees that the production level has risen to $275,000, the overhead rate should be adjusted to 70% in order to reduce the volume variance to zero. This approach should stop production management from attempting to create profits by building inventories.

In short, some variances can be ignored without harm to the company's overall cost control system by creating compensating controls and strengthening other existing controls.

COST/BENEFIT ANALYSIS

It is difficult to construct a clear-cut statement of costs and benefits for the changes advocated in this chapter, but a number of justifications for the changes are explained here. The reader can use these justifications to construct a cost/benefit analysis; however, the analysis should include a range of values based on worst/expected/best cases for each item.

Cost of Less Review of the Labor Efficiency Variance

This is the cost of having an excessive amount of staff on hand in relation to the production level. This cost is difficult to determine, since proper staffing levels are difficult to maintain in the short term, and there is a cost associated with constantly firing and rehiring employees to exactly match the level of production.

Cost of Less Review of the Labor Rate Variance

This is the cost of having too many highly paid employees in the work force in relation to the skill level required to manufacture products. This cost is difficult to determine. A correct mix is difficult to maintain, since existing employees will become more experienced (and therefore more highly paid) with time.

Cost of Less Review of the Materials Usage Variance

This is the cost of too much scrapped material. This cost can be estimated based on current scrap levels and expected changes in that level when the variance review is decreased or eliminated.

Benefit of More Review of Materials Prices

This is the benefit of ensuring that suppliers meet their contracted pricing obligations. The benefit is estimated based on expected percentages of price overages that are discovered and corrected.

Benefit of Reduced Cost of Data Recording

This is the benefit associated with not requiring employees to record variance data for labor and materials. The benefit is estimated based on the time previously needed to record the variance information.

Benefit of Reduced Cost of Variance Analysis

This is the benefit associated with not requiring the cost accountant to spend time reviewing some variances. The benefit is estimated based on the reduced number of hours needed by the cost accountant.

A typical cost/benefit analysis for revising the cost accounting system is shown in Table 5.2. It includes costs and benefits for three cases—worst, expected, and best. Some of these costs simply reduce the time requirements of personnel who will still be employed after the conversion to the revised costing system; thus, their full cost will still be paid by the company. An example is the reduced cost of variance analysis: the cost accountant will no longer be spending

TABLE 5.2 Cost/Benefit Analysis for a Revised Cost Accounting System

	Worst Case	Expected Case	Best Case
Cost of less review of labor efficiency (Assumes efficiency reductions of 10%, 5%, and 3%)	(50,000)	(15,000)	(5,000)
Cost of less review of labor rate (Assumes rate increases of 5%, 3%, and 1%)	(23,000)	(8,000)	(2,000)
Cost of less review of materials usage (Assumes increased scrap of 19%, 5%, and 2%)	(62,000)	(11,000)	(3,000)
Benefit of more review of materials prices (Assumes decreased prices of 1%, 4%, and 5%)	10,000	79,000	98,000
Benefit of reduced cost of data recording (Assumes decreased labor costs of 2%, 3%, and 3.25%)	28,000	37,000	41,000
Benefit of reduced cost of variance analysis (Assumes decreased cost accountant time of 40%, 40%, and 40%)	17,000	17,000	17,000
Net (Cost)/Benefit	(80,000)	99,000	146,000

time on certain variance reviews but will still be working for the company on new projects, so the true cost savings to the company will be zero.

In summary, it is difficult to construct cost/benefit models for changes in variance reporting, but using a risk table based on worst, expected, and best case results allows the cost accountant to construct a range of potential costs and benefits for changes to the cost accounting system.

REPORTS

The reports created by a cost accountant vary considerably, depending on the types of variances being tracked. Under the traditional system, the cost accountant is most concerned with labor rate and labor efficiency variances as well as with the spending, efficiency, and volume variances for overhead, and usage and price variances for materials. Table 5.3 summarizes these types of variances.

The traditional cost accounting variance report seems clear enough—it shows how the original variance was derived and then categorizes all the variances that roll up into the total variance. However, cost accounting is not about presenting neatly laid-out reports but about managing the reasons for costing problems. If

TABLE 5.3   Traditional Cost Accounting Variance Report

Derivation of Total Variance	
Beginning cost of goods sold	$1,000,000
+ Purchases	300,000
− Shipments	150,000
− Ending inventory balance	825,000
= Cost of goods sold	325,000
− Cost of goods sold at standard costs	280,000
= Total variance	$ 45,000
Explanation of Total Variance	
Labor variance	
Labor rate variance	$ 3,000
Labor efficiency variance	1,500
Overhead variance	
Overhead spending variance	6,500
Overhead efficiency variance	4,500
Overhead volume variance	7,500
Materials variance	
Materials usage variance	9,000
Materials price variance	7,000
Explained variance	39,000
Unexplained variance	6,000
Total all variance	$45,000

one reviews the traditional report in terms of how it helps control production costs, one uncovers the following problems.

Labor Rate Variance

Corporate labor costs tend to be fixed in the short run. So, on a daily or even monthly basis, the production manager cannot use the labor rate variance in any practical way, except to alter the mix of staff with differing pay levels on different production projects.

Labor Efficiency Variance

Efficiency variances are not of much use to the production manager, because they do not "drill down" to the next level of variance detail. At the next level, the cost accountant must review the accuracy of the labor routings used to measure the efficiency of the production staff and determine why the production staff is not able to match its efficiency goals. Too often, the problem is related to inaccurate labor routings.

Overhead Spending Variance

The production manager needs to know the details about the overhead being charged against products but frequently has no control over the contents of the overhead account. For example, the production manager is not responsible for utilities, rent, or building maintenance, even though all these items are included in overhead.

Overhead Efficiency Variance

This variance is concerned with the effect of excessive production hours being incurred; when this happens, an excessive amount of overhead is charged against production. Of course, the production manager should be concerned about the efficiency of the production staff, but if the company concentrates on producing just enough product to meet demand (a just-in-time system), it will have negative labor efficiency variances when the production staff meets its goal and stops producing. Also, if the labor force is extremely inefficient, resulting in many excess hours and therefore too much overhead charged to products, why should the cost accountant want to use labor hours as an allocation base? This leads to the final point: an activity-based costing system uses many allocation bases rather than just the labor base, which tends to result in fewer efficiency variances.

Overhead Volume Variance

This variance is concerned with the effect of an excessive number of units being produced. Since overhead is charged to the total number of labor hours incurred, a larger production volume (containing more than the usual number of labor hours) leads to more overhead charged to production. However, if the amount of overhead charged to production begins to exceed the amount of overhead incurred, then the overhead application rate should be changed to bring the amount of capitalized overhead in line with the amount of actual overhead. Since the overhead rate can be changed at will, the variance has no particular value besides signaling a need for a new overhead rate.

Materials Usage Variance

This variance is of some use, since it gives notice that there is a problem in one of several underlying areas. For example, a materials usage variance can mean that the bills of materials are incorrect and that supposedly excess usage is merely caused by parts not being included on the bills of materials that were used to formulate the costing standard. Alternatively, a usage variance could be caused by incorrect beginning or ending inventory balances, or by incorrect additions to or withdrawals from physical inventory balances. A more appropriate way for a cost accountant to monitor these issues is to track measures that directly address the problem: use inventory accuracy audits that examine inventory quantities, locations, units of measure, and valuations. Also, the cost accountant should initiate bills of materials audits that review accuracy based on the component part numbers, descriptions, quantities, and units of measure. These measures are much better than the materials usage variance because they monitor the key areas that contribute to the variance; when properly monitored and acted upon, they can even be used to keep variances from occurring.

Materials Price Variance

This is a useful variance. Since labor and overhead costs are relatively fixed in the short run, the only truly variable element left in the costing equation is the cost of materials. Even in this area, the cost accountant tends not to delve far enough into the reasons for cost increases. The production manager needs to know which specific materials costs are being increased, and by which suppliers, so that targeted pricing negotiations can be conducted with suppliers to control those costs.

Table 5.4 presents different information that the cost accountant can report on at period-end. This information is tailored to production personnel who need information they can act on, and it is presented with historical information that provides a baseline against which current information can be measured.

Average Labor Rate

This rate gives the production manager a direct indication of changes in the cost of personnel. Significant changes can be reviewed with a more detailed report that lists individual labor rates.

Average Overtime Percentage

This percentage tells the production manager if staffing levels should be changed to meet an optimal overtime target (though this decision should be made with a full knowledge of the projected production schedule).

Production Supplies, Salaries

Only the controllable elements of overhead are reported to the production staff. The rest of the overhead does not need to be reported, or could be reported as a single line item, since the production management has no control over it.

TABLE 5.4   Improved Cost Accounting Variance Report

	This Month	Last Month	Last Year
Labor Information			
Average labor rate	$12.73	$12.71	$10.21
Average overtime percent	8%	12%	4%
Overhead Information			
Production supplies	$ 3,451	$ 3,200	$ 6,215
Supervisory salaries	$82,400	$81,300	$71,000
Purchasing salaries	$38,000	$38,000	$36,000
Production control salaries	$34,000	$33,500	$33,500
Percent of overhead capitalized	92%	93%	97%
Materials Information			
Inventory accuracy percent	94%	98%	97%
Bill of materials accuracy percent	92%	99%	99%
Labor routing accuracy percent	90%	98%	96%
Variation of actual purchased parts costs from targeted costs	102%	105%	107%
Working Capital Information			
Inventory turnover	5:1	6:1	8:1

Percent of Overhead Capitalized

If this amount varies greatly from 100% over an extended time period, the overhead rate should be altered to match actual to projected overhead absorption.

Inventory Accuracy Percent

The inventory accuracy is a key measure, since it indicates the accuracy of inventory records that make up the physical inventory balance, which in turn are used to calculate the cost of goods sold. If the inventory accuracy drops below 95% in an MRP system, the system will produce inaccurate ordering information.

Bill of Materials Accuracy Percent

This is perhaps the most important measure of all. It is used to derive the list of items to be purchased, so any errors in the bills of materials will result in over- or under-purchasing. Excess purchases must be stored or returned, while under-purchasing results in rush shipments of parts at high delivery charges or in delayed production.

Labor Routing Accuracy Percent

This percentage is used to determine the amount of labor required to build a product. If the routing is inaccurate, then the size and mix of the production crew may be insufficient or excessive for the production volumes needed.

Variation of Actual Purchased Parts
Costs from Targeted Costs

Just as important as bill of materials accuracy, this measure tells the production management how closely it is matching its targeted costs. Further detail is nearly always required, which notes which suppliers are not meeting their targeted parts prices.

Inventory Turnover

If the inventory turnover drops, then the cost accountant should investigate the inventory to see if obsolescence is becoming an issue or if the production manager is building an excessive inventory to capitalize overhead costs and "manufacture" a profit.

In summary, a revised cost accounting system results in reports that differ from more traditional reports. A revised system focuses on the continuing accuracy of key data rather than investigating problems after the fact. For example, a revised system reports on inventory accuracy, whereas a traditional system focuses on the period-end materials variance. Also, a revised system reports on information that tends to be at least one level closer to the basic problem than the information provided in a traditional costing report. For example, a revised system reports on the average percentage of overtime worked per person, whereas a traditional system will only report that the cost of labor has increased, requiring the cost accountant to delve deeper to find out that increased overtime was the cause. These reporting changes allow the cost accountant and other report users to find and correct cost-related problems more quickly than would be the case with traditional cost accounting reports.

TECHNOLOGY ISSUES

Several technology issues affect the cost accountant's efficiency, particularly with regard to the accuracy of the information used to perform variance analysis. Though all these technologies have been available for a number of years, most companies do not yet use them.

Electronic Data Interchange

A company can install electronic data interchange (EDI). This technology allows a company to send electronic transactions to its trading partners, usually through a third-party computer service called a value-added network (VAN). A company can use EDI to receive period-end payables by electronic transmission. This allows the company to avoid the period-end mail float, which allows it to close the accounts payable function much more quickly and arrive at the period-end cost of goods sold figure. However, the benefit of EDI is reduced if transactions are only accessed from a VAN at irregular intervals, which sometimes occurs when companies are only sporadic users of the service.

Bar Codes

For up-to-the-minute variance analysis, a highly accurate perpetual inventory system is a must, because beginning or ending inventory balances are used to find many cost variances. The speed with which inventory transactions update the main database will determine how current the inventory records are for variance analysis. The effect of technology on this area is considerable because of the use of bar codes. Bar codes on inventory items can be quickly scanned with perfect accuracy, so that inventory transaction data can be sent to the inventory database with no need for review of keypunching errors. However, this information may still be incorrect through human error if the bar code was created with the wrong information. The most common error of this kind is when a bar code is created for an incorrect quantity. This can perpetuate errors in the system, since the operator of the bar-code scanner generally assumes that the information contained in the bar code is correct and does not cross-check the bar-coded information.

Wireless Transmissions

Wireless transmissions of inventory transactions to the central inventory database (generally linked to a bar-code scanner) can update the database instantly and give the cost accountant real-time levels of accuracy. These systems are expensive, so the cost accountant must offset the benefit of having records updated a few minutes sooner against the cost of the transmission equipment.

Just-in-Time Systems

Though sometimes construed as a management technique instead of a technological advance, the JIT system can greatly assist the cost accountant. In its simplest form, a JIT system drastically reduces or eliminates the amount of work-in-process and raw materials by shrinking work queues and by ordering materials only when required and in the exact amounts needed. The tracking of WIP is one of the largest headaches for the cost accountant, and by reducing the quantity of WIP, the variance reporting chores related to WIP are also reduced.

MRP and MRP II

Though considered old technology, a materials requirements planning (MRP) system or a manufacturing resources planning (MRP II) system can be of great use to the cost accountant, because a properly managed system will enforce the maintenance of accurate inventory and bill of materials records (for an MRP system) as well as labor routing records (for an MRP II system). This information must be accurate, because an MRP system takes production information from a master schedule, splits each production quantity into its constituent parts, and checks the existing inventory for those parts in order to place requisitions for missing materials. If the inventory or bill of materials information is inaccurate, the MRP system will not function properly. By extending this system into the units of work required to create products (as defined in labor routings), the system can predict capacity problems in the production facility (this added functionality is called MRP II). Once again, if the labor routing information is inaccurate, then the system will generate inaccurate results.

All three of these databases (inventory, bills of materials, and labor routings) are used by the cost accountant to determine the causes of costing variances in areas related to materials and labor. In short, either kind of MRP system introduces more accurate information into the variance analysis and costing processes.

All the preceding technological advances assist the cost accountant by improving the speed or accuracy of transactions, resulting in fewer variances to investigate.

MEASURING THE SYSTEM

Table 5.5 describes the formulas used to track the key information for a revised cost accounting system, shown in Table 5.4. It does not include those variances used under an old-style costing system, since it is assumed that the reader already has access to the formulas for the various kinds of materials, labor, and overhead variances.

TABLE 5.5 Formulas for Key Elements of Revised Cost Accounting System

Element	Formula	Comment
Average labor rate	Total labor dollars expended in period ÷ Total hours worked in period	Can be broken down by department to show changes in rate for each manager's area.
Average overtime percent	Total hours worked in period ÷ Total regular hours worked in period	Can be broken down by department to show changes in percentage for each manager's area.
Overhead information	—	Includes only costs that are controllable by manufacturing management. Usually includes supplies and salaries related to manufacturing and materials management.
Percent of overhead capitalized	Total actual overhead cost ÷ Total capitalized overhead cost	Can also be shown as the dollar amount under- or over-absorbed.
Inventory accuracy percent	(Total parts counted − Items with incorrect locations, quantities, descriptions, or units of measure) ÷ Total parts counted	—
Bill of materials accuracy percent	(Total bills reviewed − Items with incorrect parts, quantities, or units of measure) ÷ Total bills reviewed	—
Labor routing accuracy percent	(Total routings reviewed − Items with incorrect times, machines used, or employee types) ÷ Total routings reviewed	—
Variation of actual purchased parts costs from targeted costs	Total cost of purchased parts ÷ Total cost of purchased parts as per contracts or P.O.s	Can be shown by supplier to highlight suppliers who consistently overcharge for materials shipped. Can also be shown by number of parts with incorrect prices rather than by total variation for all purchased parts.
Inventory turnover	Cost of goods sold ÷ Average inventory	Can be broken down into turnover for raw materials, work-in-process, and finished goods, so that slow turnover areas can be more easily seen.

The controller needs to use a different set of ratios when evaluating the performance of a JIT manufacturing system. A JIT system operates on the principle that the facility should only receive enough supplier components to build parts, produce only enough parts to build the desired number of products, and only produce enough products to meet demand. In order to produce with the exact number of required components from suppliers, the receipts must be delivered to the company on time, in the right quantities, and with perfect quality (no defective components). In order to produce only enough parts to build the desired number of products, setup times must be minimized, work-in-process must be drastically reduced, and scrap must be carefully tracked. In short, the controller must devise data collection procedures for information that does not appear on the balance sheet or income statement. The only ratio related to JIT that can be derived from the balance sheet is inventory turnover.

In short, a variety of alternative measure are available that provide more relevant costing information to management than the traditional set of volume and efficiency variances.

IMPLEMENTATION CONSIDERATIONS

Removing some of the traditional cost accounting variances can be a traumatic experience for those manufacturing executives who have become used to reviewing them every month for the last few decades. One of the easiest transition techniques for these executives is to provide both the old and new variance information until they are comfortable with the new information. However, the accounting executive implementing the changes should specify a date when the old variances will no longer be provided—otherwise there is a danger that the cost accountant will be burdened with providing both sets of information in perpetuity, which not only increases the accounting work load but also slows down the speed of period closings.

Also, cost/benefit analysis can be used to convince manufacturing executives that they can avoid the costs associated with data collection for variance analysis as well as free the cost accountant to review materials price variances. Any reduction in materials costs makes the manufacturing department look better, so this should be a welcome improvement.

SUMMARY

This chapter has described how the traditional cost accounting position no longer provides information to management that is current or relevant. A new set of costing information is available that could be more useful to management. At the same time, most of the new information to be reviewed by the cost accountant falls outside of the normal period closing process, so the speed of the accounting close can actually be improved while more relevant information is being provided. The chapter also notes a new reporting format for issuing this information to management as well as a discussion of control issues regarding abandonment of some cost accounting variances. In short, this chapter presented a win-win situation where the provision of superior costing information to management actually improves the speed of accounting transactions.

REFERENCE

Hackett Group. *AICPA/THG Benchmark Study: Results Update and Analysis.* 1994.

6

PAYROLL

In most companies, payroll processing requires a large staff of clerks to enter a variety of changes to employee pay rates and deductions. A fairly standard program then calculates pay for employees, and paychecks are distributed. Even a cursory view of the function reveals that the data input part of the payroll process is by far its most inefficient part. While making some comments about other parts of the payroll function, this chapter concentrates on how to improve the data entry portion of payroll processing, so that fewer employees are required, data entry errors are decreased, and the payroll can be processed more rapidly than before.

CURRENT SYSTEM

This section breaks the payroll processing function into three steps and describes the processes and controls used in each.

Payroll Data Input

The payroll data input step enters information from a variety of sources into the payroll software for later calculation. The information is typically entered by a team of payroll clerks. The most common types of information entered are as follows.

Employee Hours Worked. At the end of each pay period, the accounting staff enters the hours worked for each hourly employee into the payroll system. A variation is to enter piece rates for each employee. The entries must factor in the amount of vacation or sick time taken by each employee as well as any overtime payments. Also, depending on company policy, employees will not be paid for various fractions of an hour if they arrive a few minutes late for work. To make matters even more complicated, some companies who charge costs to specific jobs must also enter the time worked for specific job numbers—this information is passed to the job-costing system for summarizing into total costs incurred for individual jobs. In short, the most time-consuming and error-prone payroll data entry task is the one involving entry of hours worked.

Bonuses and Commissions.   At irregular intervals, the accounting staff calculates commissions based on such variables as a percentage of revenues, splits with other salespeople, splits with sales supervisors, cumulative year-to-date bonuses, regional premiums, and percentages of gross margins on products sold. Commissions may be paid based on cash received, revenue booked, or cash received less bad debts. Bonuses can be based on similar measures. In short, these calculations can be difficult to track and summarize, and the number of methods used increases with the ingenuity of the management team in coming up with new bonus and commission payment schemes.

Benefits Deductions.   Many companies offer employees a variety of dental, health, life insurance, and other benefit plans to choose from, each of which has a deductible that must be removed from the employees' paychecks. Since deductibles may change based on which set of benefit options is chosen, this can be a difficult deduction to track. Also, the paperwork is typically filled out by the employee and passed to the benefits department, which in turn gives the deduction information to the payroll clerks for entry into the computer. Since there are a number of people involved in the process, it is easy to enter an incorrect deduction amount.

Tax Deductions.   Employees request changes to their deduction amounts from time to time for both federal and state withholdings. These changes may be a percentage of gross earnings or a flat amount. In either case, the payroll clerk makes the entry based on signed paperwork filled out and submitted by employees.

Pension Plan Deductions.   Some pension plans, such as the 401(k), require the employee to contribute either a percentage of gross earnings or a flat amount to a fund, to which the employer also contributes. Though some plans allow only biannual changes, the majority are more frequent. This means that the payroll clerks may be called upon to enter a large number of pension contribution deduction changes per year.

Garnishment Deductions.   Companies may be required by law to deduct specific amounts from an employee's paycheck to cover court-ordered payments resulting from a legal settlement. This requires not only the initial entry by the payroll clerk of the garnishment amount but also tracking the recurring garnishment deduction until the deduction terminates.

Distributions to Bank Accounts. If the company uses direct deposit to send payroll payments directly into employee bank accounts, then the payroll staff must enter the bank number and account number into the payroll system. In addition, employees may request multiple distributions, so that a single paycheck may be split up into deposits that are sent to mutual funds, savings accounts, and checking accounts. All these accounts require a setup entry by the payroll clerks.

Shift Differentials. Many companies pay employees a premium to work on the second or third shifts, since these are less desirable times to work. The payroll staff must enter these shift differential premiums into the payroll system every time an employee changes shifts, though most systems will automatically pay the premium once an employee has been set up to receive the payment.

Personal Item Deductions. Many companies allow their employees to purchase items through the company, since it can command reduced prices on a variety of products. The employees then pay the company through payroll deductions. If the item purchased is quite expensive, there may be a series of deductions that span many payroll cycles. The payroll staff must enter these deductions and track the amount withheld until the full amount has been paid.

Pay Rate Changes. With supervisory approval, the payroll staff must enter pay changes for staff members. These can be more frequent than once a year, since new employees frequently receive a review after 90 days with the company, and pay raises are given with most promotions.

New Employee Additions and Departing Employee Deletions. With proper supervisory approval, new employees must be added to the database. These entries may require such additional information as emergency contact names and phone numbers, social security numbers, mailing addresses, start dates, and initial pay rates. Also with proper supervisory approval, the payroll staff may delete departing employees from the payroll system. In both cases, the payroll staff must calculate partial payments for partial periods worked, since employees do not always conveniently begin or stop work at the beginning or end of the company's payroll cycle.

Controls are needed when inputting payroll information. In particular, pay changes, new hires, and employee deletions are not input without a signed authorization from management. Also, hourly employee time cards are usually reviewed in advance by the supervisor of the hourly employees for any irregularities and signed before being passed on to the payroll clerk. Bonuses and commission calculations also are normally reviewed by a supervisor prior to payment. In general, any change to an employee's gross pay requires advance approval by management. However, deductions from an employee's pay (e.g., 401(k) deductions, benefits deductions) only require the approval of the employee. This is a key control issue that allows the company to automate key features of the payroll data input step.

Processing

The processing step takes all the information entered in the previous step to calculate gross wages from hours worked, commissions, and bonuses. Deductions from the gross wages earned include taxes, benefits payments, pension plan contributions, garnishments, shift differentials, and payments for personal items. These deductions result in a net payment to each employee. The system then divides the net payment into payments by check or by direct deposit to various bank accounts.

The processing step is usually carried out by off-the-shelf software. The calculations are so routine that it makes little sense for most companies to create a customized software package. Also, off-the-shelf packages have been thoroughly tested, so that programming bugs are less likely to disrupt the payroll calculation process—a late paycheck does not improve the employees' morale. An increasingly common variation is to transmit the payroll information to an outside payroll processing service, which calculates gross pay, deductions, and net pay, and delivers printed checks back to the company.

For companies with subsidiaries, payroll systems usually are not integrated, so there is at least one payroll system per location. This can cause problems when employees rotate among subsidiaries with any regularity, since they are set up as new employees on the payroll system of the subsidiary they are moving to, even though they may have considerable tenure with the corporate parent. This creates problems when tracking years of service with the company, since this information is needed when applying vesting percentages to corporate pension plans.

Controls over the processing step primarily relate to access to the program. Thus, the control is simple—access is restricted to specific user identification codes and passwords. From the perspective of the computer systems manager, it is wise to retain multiple off-site copies of the payroll database and application program in case of damage to the primary computer storage facility. This backup feature is also useful when an update to the payroll package has software glitches—the previous version can be restored to the computer from a backup storage medium so that the payroll can still be processed.

Distribution of Payments

The distribution of payments step involves printing checks, signing them, and distributing the checks to employees as follows.

Set up Printer. Check stock is removed from a secure location and set up on a printer. The check stock nearly always includes a remittance advice, on which is printed gross pay, deduction, and net pay information.

Align Printer. A set of voided checks are printed to position the checks on the printer. The voided checks are discarded.

Print Checks. The checks are printed.

Sign Checks. The checks are either signed automatically with a signature plate or signed manually by an authorized check signer.

Enclose Checks. Checks are sealed in envelopes with the employee name either appearing through a window or being printed onto the envelope.

Deliver Checks. An employee designated for the task (sometimes called the paymaster) distributes all checks into the hands of the employees to whom the checks are addressed. Excess checks are secured in a safe and distributed when the employee can be found. This creates problems for salespeople and other employees who travel and who are not necessarily in the office on payday to receive their checks.

A primary control over the distribution of checks is control over the check stock. If a blank check is stolen, a payment amount and signature can be forged onto it. The obvious control is to store the check stock in a locked location and restrict access to the key. In addition, the payroll staff maintains a log of check numbers used; this log is useful for identifying the absence of checks.

Another control is the signing of checks. Supervisors who sign checks can determine, based on their own knowledge of who works for the company, if any checks are being cut for people who no longer work for the company or for amounts that are clearly excessive. If the company has grown so large that the signer cannot know everything about all employees, this is still a good control for a different reason—the signature makes the check an official commitment to pay the employee, whereas a stolen, unsigned check has no such validation. If the company views the control in this respect, then a signature plate can be used instead of a supervisor to print a signature. Of course, the signature plate must be locked in a secure location when not in use.

Another control is over the delivery of checks. Many companies use a paymaster, an employee specifically designated to take the checks from the accounting department and give them to the employees. The paymaster must account for any checks that were not handed to employees. Checks should only be given to the employees to whom the checks are addressed rather than to other employees who will deliver the checks; when other employees have the checks, they can attempt to cash the checks. Also, since an employee must be on site to receive a check, any checks created for nonexistent employees will never be issued. A final control over the disbursement of payroll checks is that the paymaster not be involved in the processing of payroll. This control keeps a payroll clerk from creating a check and then disbursing it to himself. A flowchart of the standard payroll process is shown in Figure 6.1.

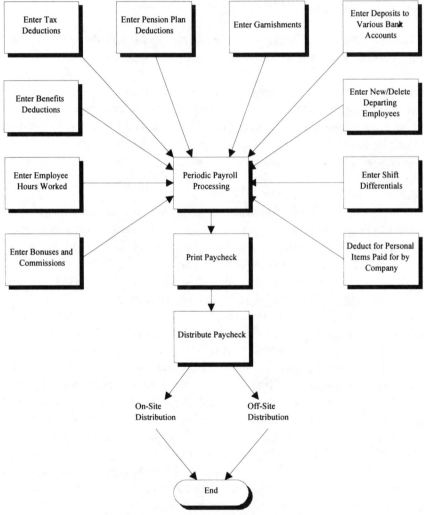

FIGURE 6.1 Standard payroll process.

There are also a number of government-mandated payroll reports to be issued on a quarterly or annual basis. The most time-sensitive report is the W-2 form for each employee, which is issued following the end of the calendar year.

A review of Figure 6.1 and the discussion of the payroll process reveals a great deal of movement of paperwork among personnel. The move and wait times thus introduced into the process greatly slow it down, and the potential exists for the loss or misinterpretation of the paperwork. A payroll value-added analysis is shown in Table 6.1. A value-added item is considered to be one that brings the payroll transaction closer to conclusion. The time estimates in the table are based on a single working day of activity, not a full week of time recording.

Table 6.2 shows that only 15% of the steps bring the payroll transaction closer to conclusion; the remaining steps are related to moving paperwork from person to person or reentering information that has already been completed by the employee. In terms of time required, the value-added steps can be concluded in one-third of an hour, whereas the non-value-added and move portions of the transaction take up twice as much time. The payroll processing function is unusual in that there is no wait time; because of the confidential nature of payroll information, it is brought straight to the payroll clerk instead of languishing in the payroll clerk's mailbox. In short, the actions needed to conclude the transaction are a small proportion of the total process.

In summary, a large amount of time is required to input wage and deduction information into the payroll software. These entries are subject to input errors, with resulting employee dissatisfaction. The payroll processing is usually automated, with few processing problems, while the paycheck-printing process is also relatively free of problems. The next section discusses ways to reduce the work load of the payroll staff in entering payroll information, while at the same time reducing the number of entry errors and increasing the transaction speed of the entire process.

REVISED SYSTEM

In many companies, the payroll process is slow to complete, requires a large staff to enter payroll-related information into the computer, and results in many data entry errors. By way of comparison, the cost of a payroll transaction in a world-class company is $1.72 versus $5.00 in a Fortune 100 company.[1] Clearly, some companies have figured out how to reduce the cost of payroll far below the norm.

---

[1] Steve Coburn, Hugh Grove, and Cynthia Fukami, "Benchmarking with ABCM," *Management Accounting,* January 1995, p. 59.

TABLE 6.1   Payroll Value-Added Analysis

Step	Activity	Time Required (Minutes)	Type of Activity
1	Employee enters time worked on time card.	5	Non-value-added
2	Payroll clerk moves to time card storage rack.	1	Move
3	Payroll clerk removes time card from storage rack.	1	Non-value-added
4	Payroll clerk returns to office.	1	Move
5	Payroll clerk reviews time card for overtime.	2	Non-value-added
6	Payroll clerk reviews time card for absences.	2	Non-value-added
7	Payroll clerk reviews time card for missing time entries.	2	Non-value-added
8	Payroll clerk enters hours worked into payroll software.	5	Non-value-added
[Payroll clerk receives benefits deduction information from human resources department.]			
9	Payroll clerk enters benefits deduction changes into the payroll software.	2	Non-value-added
[Payroll clerk receives tax deduction information from employees.]			
10	Payroll clerk enters tax deduction changes into the payroll software.	2	Non-value-added
11	Software calculates amount payable to employee.	0	Value-added
12	Clerk prints payroll checks.	10	Value-added
13	Clerk brings checks to authorized check signer.	1	Non-value-added
14	Authorized signer reviews and signs checks.	5	Non-value-added
15	Clerk stuffs checks into envelopes.	5	Non-value-added
16	Clerk brings checks to paymaster.	1	Non-value-added
17	Paymaster delivers checks to employees.	10	Value-added
18	Paymaster brings undistributed checks to accounting department.	1	Non-value-added
19	Clerk stores undistributed checks in a safe.	1	Non-value-added

This section discusses how to speed up the payroll process while cutting down on data entry labor and errors, resulting in greatly reduced costs to the company.

One of the problems with a traditional payroll system is the number of people who handle deduction information before it finally enters the payroll system. For example, an employee selects a medical plan; the human resources department calculates the employee-paid portion of the plan and forwards this amount to the

TABLE 6.2 Summary of Payroll Value-Added Analysis

Type of Activity	No. of Activities	Percentage Distribution	No. of Hours	Percentage Distribution
Value-added	3	15%	.33	35%
Wait	0	0	0	0
Move	2	11	.03	3
Non-value-added	14	74	.58	62
Total	19	100%	.94	100%

payroll department, which enters the information. The employee receives a re-mittance advice that lists the amount of the deduction, and wants to change to a cheaper alternative. This sets off another round of changes. Not only does this approach involve the participation of too many people (any one of whom could pass along incorrect information) but it may take too long to enter the informa-tion, especially if the deductions are being sent from an outlying location to a centralized payroll system. This is of considerable concern to a company, which grants a benefit to an employee as soon as the employee signs the paperwork but which cannot enter the related deduction information into the payroll system quickly enough to recoup part of the cost of the benefit. Of course, the payroll staff can always make a double deduction in the following pay period to make up for the missing deduction from the prior period, but this requires additional time by the payroll staff to make the manual entry.

The company can set up an electronic form on its computer network *giving employees direct access to their deduction records* so that they can alter their own deduction information. Since not all employees have access to a computer termi-nal, the company can provide access to a free-standing computer kiosk that al-lows access to the deductions, or it can allow access by touch-tone telephone to the same information. A combination of these three options should allow all company employees access to their deduction records. A flowchart showing di-rect entry of deduction changes by employees is presented in Figure 6.2. Employ-ees enter their employee numbers and personal identification numbers (PINs) to gain access to their deduction records. (The system does not allow access to pay rate information, since changes to that information require supervisory ap-proval.) If the deduction is a tax deduction, each employee can enter either a fixed amount or a percentage of income. To keep employees from being surprised at the net amount of their paychecks when paid, the system immediately calcu-lates the amount of net pay left and communicates this amount to the employee. If the net amount is acceptable to the employee, then the change is made. Pension deductions operate in a similar manner. Benefits, however, are slightly different. Most companies have a variety of benefits options, so the employee may need to

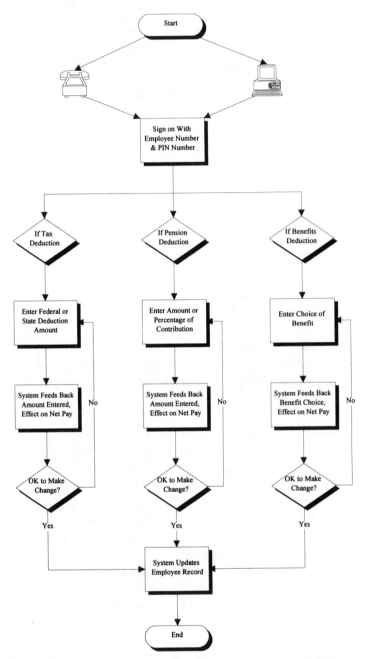

FIGURE 6.2 Direct entry of deduction changes by employees.

select a combination of several dental, medical, disability, or life insurance plan options, each of which has a different deduction amount. The effect on net pay should be communicated to the employee as each benefit is selected, so that the employee can determine the cost of individual benefits rather than only the cost of all the benefits together. In short, direct access to deduction records allows employees to enter deduction information themselves with immediate feedback, which eliminates errors and keeps the payroll data entry staff from having to enter any deduction information at all.

Another problem with the payroll process is having to make deduction entries for those employees who have purchased items through the company and who are now compensating the company for the amount of the purchase. The payroll staff must enter the amount of the deduction per pay period and track the deductions through to their termination date, when the deduction is removed from the system. This process increases the work load of the payroll clerks, not to mention the general ledger clerk, who must make a journal entry to account for the repayment. Since buying items for employees does not fit into the mission statement of any company, the best solution to this problem is *creating a policy of not allowing company purchases of merchandise for employees.* By removing the need for the deduction, the deduction-related work disappears.

Another problem is the amount of work required to calculate commissions. These calculations can be so complex that it is too expensive to generate the program code needed to automate the process, resulting in manual calculations by the payroll (or other accounting) staff. This problem is exacerbated if there are a large number of invoices to summarize by sales territory for each salesperson's commission calculation. The solution to the problem is a difficult one for the controller or CFO, who must convince the sales manager to adopt *a simpler commission structure.* The simplified structure is then automated, so that the computer system generates a list of commissions payable without any input from the accounting staff. Getting the sales manager to adopt a less complex commission structure is difficult, because the complexity of the typical commission system is designed to foster a certain type of behavior by the sales staff, and a change to that system may not result in optimal sales. However, by educating the sales manager about the time and effort required to calculate commissions, some mutual understanding of the problems of both parties (the sales and accounting departments) will result, which should yield a simplified commission structure that is satisfactory to both parties, though perhaps not the optimal solution for either department.

From the perspective of the accounting department, the ideal commission structure is one that can be automated with minimal programming effort. The following are, in declining order of ease of programming, various commission calculation strategies.

Percentage of Invoices Issued.   This is the easiest programming task, because any computerized accounting system should already have a database of invoices that can be sorted by salesperson code and date to derive the amount of commissions payable for each pay period.

Percentage of Cash Received.   This is a more difficult programming task, because invoices may not be paid for in full by the customer, resulting in tracking commissions already paid for a partial customer payment, since the company does not want to pay the salesperson again for the full amount of the invoice when the remainder of the cash payment is received from the customer.

Percentage of Invoices Issued, Less Bad Debts.   A company's bad debts are commonly kept in a manual file, not in a computer database, so the programmer must create a file for the storage of bad debt information in order to link it to the invoices file, thereby allowing bad debts to be deducted from the amount of invoices issued.

Any Preceding Method, Plus Splits with Other Salespeople.   This is a more difficult programming task than may be immediately apparent, especially if combined with payments based on a percentage of cash received. This is because there must be room in the invoice file for multiple salesperson identification codes as well as multiple commission percentage codes, which requires a change in the structure of most databases. In the case of commissions based on cash received, it also means that the computer system must track the differing commission amounts still owed to multiple salespeople on invoices that have only been partially paid.

Any Preceding Method, Plus Differing Commission Rates Depending on Gross Margins.   The most difficult task of all is creating fields in the database for the gross margin on each product sold as well as a separate table that issues a different commission rate based on the amount of the gross margin. To complicate matters further, the gross margin may be tied to actual costs, so that the gross margin may fluctuate over time. To introduce yet another programming problem that occurs when commissions are paid when cash is received, the commission table may have to store older commission rates in case salespeople are compensated based on commission rates existing at the time the product was sold rather than the amount in effect when cash was received in payment for the invoice.

The highest-volume item involved in the payroll data entry process is entering hours worked for those employees who are paid on an hourly basis. This information is usually translated from time cards and manually entered into the payroll software by the payroll department. Since the hours worked must be manually transferred, there is a high probability of errors being made by the payroll department. When errors are discovered, the payroll staff must bring in the questioned time card from storage, recalculate the time, correct the error, and file away the time card—a considerable waste of time. To improve the situation, the company can *install a bar-coded or magnetic stripe time card system.* With this system, each employee is given an employee card with an identifying bar code or magnetic stripe on the back. When an employee arrives, leaves, or takes a break, the card is run through a scanner, which automatically identifies the employee and records the time in a computer file. The scanner can also be used in concert with a keypad to punch in the time worked on specific jobs. The scanner is linked to a computer that stores the time information and that can be linked with a custom interface program to the company's payroll software, so that all employee time information can be uploaded for payroll processing without any manual entry being required. In reality, however, reviewing an edit report of time records for any obvious errors is a reasonable intermediate step prior to printing paychecks.

A time card scanning system has other benefits. For example, it can reject a scan if the employee has arrived late for work and will only allow the entry if a supervisor punches in an authorization code in advance; this keeps management informed of late arrivals. Also, the system can automatically enter an employee's missing time entry if she forgets to clock out. Another feature is not allowing time entries for shifts in which an employee is not supposed to be working, thereby keeping an employee from punching in for the wrong shift and collecting a pay differential for the shift. Finally, management can print a variety of reports from the system, such as a listing of who has not shown up for work, who is chronically late, and who forgets to clock in or out. In short, there are a variety of reasons why an automated time keeping system should be installed.

Another problem with payroll systems is the large number of payroll-related databases that are not linked. For example, the payroll database typically contains an employee's name, address, social security number, deductions, and pay information for the current year. The benefits database is usually kept manually and includes an employee's name, address, social security number, and listing of current benefits selected. Finally, the human resources department maintains a database of an employee's name, address, social security number, contact information, pay history, job classification, and history of work-related accidents. When maintained separately, there is a strong likelihood of varying addresses, deductions, and even pay amounts being stored, since not all the data are updated when a change is made—to do that, changes may have to be made as many as three times to cover the three sets of data. It is not efficient for the accounting and payroll staffs to have to cross-check each other's databases to find out which information is correct; this reduces the speed of the entire payroll process. The obvious solution is *combining the three databases into one database* (see Figure 6.3). With one database, only one update is required to change information. This reduces the time spent by the accounting and human resources departments in updating employee information.

Along similar lines, the controller of a larger firm should try *reducing the number of payroll systems.* One study found that the average company has 3.6 payroll systems per billion dollars of revenue. This is frequently because there

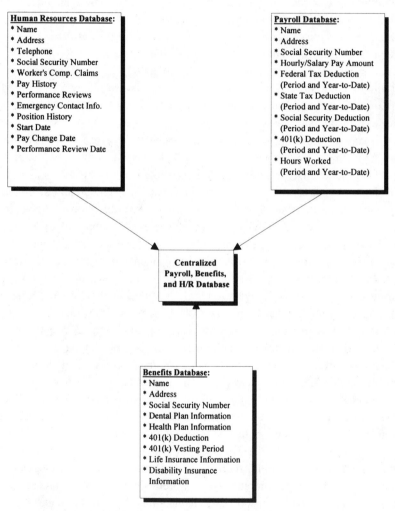

FIGURE 6.3   Consolidating payroll databases.

are different systems by geographical location. However, it is not a good situation to have if centralized payroll information is needed for reporting on overall company pay rates, benefits, and other information needed by top management. Merging systems requires centralization in a single location but may allow the company to set up linked terminals in all company locations so that local staff people can enter payroll changes, while the payroll database is located centrally, with payments being made from the central location.

Turning to the output of information from the payroll process, the company creates a check that it gives to each employee. This creates a problem for employees who are not on hand on payday to receive their checks. This is also a problem for centralized payroll systems, since the company must use an expensive overnight delivery service to send paychecks to outlying locations in time for payday. In addition, by insisting on manually delivering checks to employees (and there are good control reasons for doing so), the company must pay an employee to make the distribution. A cleaner approach is *using direct deposit to pay employees.* With this method, pay amounts are automatically deposited in employee accounts on payday. This eliminates the need for a manual distribution of paychecks and allows employees to receive their pay even when they are not on hand to receive a paycheck on payday. Since some employees do not have bank accounts in which funds can be deposited, the company can team up with a local bank (or several banks, in case of multiple plant locations) to offer free bank accounts to those employees without accounts. The company then uses direct deposit to send money to the new accounts. It is easy to find banks who are willing to extend this service, since they gain new depositors and the temporary use of the money deposited in the bank accounts (until it is withdrawn).

It is still a good idea to give employees a remittance advice that lists their gross pay, deductions, and net pay, but as long as the money reaches employees on time, there is no need for the remittance advice to be given to employees at the same time—instead, cheaper postal service rates can be used to send remittance advices to outlying company locations a few days after the pay date. An alternative to sending a printed remittance advice is linking the payroll system to the company's electronic mail system, and *sending the advices as electronic mail messages to employees.* This eliminates the cost of paper and postage when sending out remittances but is only good for those employees who have electronic mailboxes—other employees would still have to receive the paper version of the remittance.

A final issue that slows down the functions of the payroll staff is responding to inquiries by employees about the amount of vacation and sick time they have remaining. This is the most common request from any company's staff. It slows down the payroll department by taking clerks away from other tasks to calculate the most recent accrual of vacation or sick time and communicate that information to the employee. A simple way to avoid this problem is *including the amount of the accrual on the payroll remittance advice* that goes to each employee on payday. This is quite effective for weekly pay periods, though somewhat less effective if the company has a one-month pay period, since accruals and vacation and sick time usage may vary considerably through the month. In the latter case, the company can provide the information to employees through the company's network, over the telephone, or through free-standing computer terminal kiosks. The net effect of this change is that more time is available to the payroll staff to complete payroll processing.

In summary, a variety of techniques, nearly all of them involving some programming work, can be implemented to increase the speed of the payroll process while reducing the number of payroll-related errors. By merging databases, automating commission calculations, allowing employees to enter their own deductions, using automated time recording, altering the paycheck remittance advice, and using direct deposit, the ease of payroll processing can be improved. A flowchart of the revised payroll process is shown in Figure 6.4. Some of the controls related to payroll are altered or rendered ineffective by these changes. The next section discusses alternative control systems that allow the controller to maintain tight control over the more streamlined payroll process.

CONTROL ISSUES

There are a number of control issues related to payroll. For example, if the controller were to install a bar-coded time-keeping system, employees could make a copy of the bar code on the back of the time card and have someone else scan the copied bar code into the scanner at any hour, resulting in more hours earned. As another example, use of direct deposit to pay employees makes it more difficult to spot payments to nonexistent employees. Further, sending payroll remittances to employees via electronic mail means that employees do not even have to have a mailing address to receive a check, which makes it even easier to pay a bogus employee. This section reviews several control problems that arise as a result of the streamlining methods noted earlier, along with possible solutions.

Employees Can Access Deduction Records

Having payroll clerks enter deduction changes into the computer database from signed deduction request forms keeps deductions from being illegally altered by people other than the employees requesting the change. If the control were removed, anyone could enter the deduction system from a computer terminal or a telephone and change key deduction or benefit amounts.

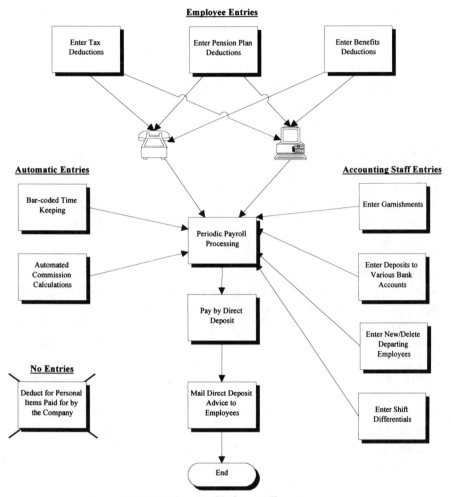

FIGURE 6.4   Revised payroll process.

Use Access Codes.   All employees must use identification codes and PINs to enter the deduction system, thereby restricting access to those people who know the codes (which may not be just the employees, depending on how well employees are protecting their access codes). More protection can be added by having the computer system automatically create new PINs at regular intervals, which can be communicated to employees by sealed envelope.

Use Video Cameras on Kiosks.   If unauthorized parties are regularly altering the deduction records of employees, then access to the deduction records can be limited to a small number of computer terminal kiosks, which can then be fitted with video cameras to record the faces of everyone making transaction entries. The mere existence of the cameras will deter most people from making unauthorized changes, though restricting access to these kiosks does reduce the ability of employees at outlying locations (or traveling employees) to make deduction changes.

Employees Can Copy Time Cards

Having employees use cardboard time cards to record their hours worked by punching into a time clock is a good control, because a permanent visual record is kept of the exact time worked by each employee that is difficult to forge. When a bar-coded time card is used, the bar code may be copied and used to enter additional time for an employee; alternatively, the time card may be given to another employee, who can check in an employee who is actually absent.

Protect Bar Codes with Red Laminate.   Bar code badges can be covered with a red laminate that prevents copying machines from creating a copy of the bar code under the laminate, thereby preventing multiple copies of badges from appearing in the workplace (and being used to check in employees who are not there).

Print List of Employees on Site.   Nearly all time card scanning systems also allow supervisors to print reports listing the employees who, according to their entries on the scanning machine, are currently working in the production facility. This report is useful for determining the actual physical presence of employees. Any employees listed on the report who are not present can have their time cards deactivated, so that they must see a supervisor before they can clock in again.

Direct Deposit Does Not Allow
Direct Payment of Employees

When employee pay is restricted to checks that are physically handed to the employees, it is easy to determine if checks may have been printed for employees who do not exist, since they are not on hand to receive a payment. When direct deposit is used, this control disappears, because the payment is sent to a bank account instead of being handed to the employee. This may result in payments being made to fictitious people or to employees who are no longer employees but who were never removed from the payroll system.

Distribute Remittance by Hand.   If the payment is sent to the bank, the remittance can still be given to employees by the paymaster, thereby ensuring after the fact that each paid employee does exist. This means that the company will lose one direct deposit that has already been made to a fictitious employee but will not make the mistake again after the paymaster discovers that the employee does not exist. Even the first deposit can be prevented by having the paymaster hand out remittances before the date on which the deposits are set to be sent to employee bank accounts, thereby giving the company time to cancel a deposit. However, continuing to hand out *any* paper-based material wastes company resources. Instead, the next control point should be adopted as the preferable measure.

Strengthen Previous Controls.   One might ask why the company cares about determining why employees exist *after* the paycheck has already been printed. Any need for reviewing employee existence once paychecks have been printed merely assumes that the controls have failed. The correct approach is to strengthen prior controls. Those controls include a continuing audit of employee existence (probably conducted periodically by the internal audit department), supervisory review of hourly time records prior to check printing, and the prompt entry of information into the payroll system about employees leaving the company.

Electronic Mail Does Not Allow Direct Payment of Employees

When an employee receives a payroll remittance advice by hand or at least through the mail, the company has a way to track the existence of the employee, even if the pay has already been sent to a bank account by direct deposit. However, when remittances are sent by electronic mail, the company has no way of knowing if there is even a mailing address, since an electronic mail address can be independent of a mailing address. This can result in payments being made to fictitious employees or to people who have left the company but were never removed from the payroll system.

Only Send Remittances to Company-Owned Electronic Mail Addresses. The company should restrict electronic remittances to company-owned mailboxes. The company can control the existence of its electronic mailboxes by immediately shutting down mailboxes when employees leave and reviewing the open mailboxes regularly (possibly a task for the internal audit department).

In summary, when new streamlining methods are implemented, some controls can be eliminated without damaging the company's overall set of payroll controls. This can be done by strengthening other existing controls or by creating new controls, such as making it impossible to copy bar-coded employee time cards. Taken together, the control changes advocated in this section will provide effective control over the new payroll system.

QUALITY ISSUES

Payroll transactions are slowed considerably by the introduction of mistakes into the process. Thus, high-quality payroll processing is defined as processing with the smallest number of errors. This section notes the most common payroll mistakes and suggests ways to keep them from happening.

Employees Enter Incorrect Deduction Amounts

If an on-line employee deduction system is installed, employees may inadvertently enter incorrect deductions, which they will not discover until they receive their paychecks. To avoid this problem, the on-line deduction system should print each employee's entry on screen and ask for verification before entering the change.

Automated Commission Calculations Overlook Special Payment Situations

If an automated commission calculation system is installed, it is easy to have payments be made to the wrong salesperson based on special splits or misapplications of invoices to the wrong territories. The controller can solve the problem either by reviewing automated commission calculation reports prior to commission checks being cut or by adding controls when invoices are first entered into the system. If a manual review is used, the controller is simply adding manual labor back into a process that was supposed to be automated, thereby eliminating the value of the automation. A better approach is to strictly simplify the application of invoices to specific sales territories, which is performed during the order entry stage—in short, the invoice should be credited to the salesperson who is responsible for a specific zip code or state. Any other options that allow commission payments to go to different salespeople based on varying criteria are probably too hard to automate.

Direct Deposit Does Not Allow a Last Review by the Check Signer

The check signer occasionally finds a mistake in payment amounts when signing paychecks. If direct deposit is used, then there is no check signer, so there may be an increased incidence of errors. To avoid this problem, the controls used prior to the check-signing stage should be strengthened. Specifically, the check edit report should be printed and closely reviewed before the direct deposit information is sent to the bank, since the edit report is a summary of all changes made to the payroll prior to payment.

Bar-Coded Time Card System Provides No Feedback to User

A common problem with a time card "swipe" system is that the user swipes the card through the reader and then immediately swipes the card again, because the reader unit gave the user no indication that the first swipe worked. This results in a user clocking himself in and then right back out again, with no time worked, or vice versa. Employees who have no recorded hours worked in the payroll system are very unhappy people. To avoid this problem, the newer scanning systems should be purchased; these units have a lighted display that shows the name of the individual who just scanned into the machine as well as the time and date of the scan. This information assures the user that the card swipe actually worked. Without some visual feedback of this sort, employees tend to do a double swipe frequently.

Employees Do Not Check Their Electronic Mail for Payroll Remittances

A very common problem is that employees do not check their electronic mail. When remittances are sent by e-mail, employees may not receive the information because they never check their mailboxes. Actually, this is a minor problem, because many employees do not check their mail because they feel that there is nothing important for them to receive. When remittances appear through e-mail, they will have a much greater motivation for accessing their mailboxes. Other ways to initiate more user access is by training all employees in the use of mailboxes when the system is installed, with repeat training from time to time. Another possibility is to have a prize program set up through electronic mail, in which messages are sent out at random—if responded to within one day, a minor prize is handed out. This keeps people checking their mailboxes for mail.

In short, there are a few ways in which errors can still creep into the improved payroll system, but installing feedback systems, keeping calculations simple, and strengthening a few preexisting controls should keep them to a minimum.

COST/BENEFIT ANALYSIS

This section shows how to write cost/benefit analyses for implementing direct access by employees to their deduction records, bar-coded time card systems, direct deposits to employee bank accounts, electronic mail deliveries of payroll remittances, and altering payroll remittance advices to include vacation and sick time accruals. In these examples, expected revenues and costs are as realistic as possible. However, to conserve space, they do not include lengthy net present value calculations.

Give Employees Direct Access to Their Deduction Records

The controller of the Circular File Company, manufacturer of wastebaskets, is concerned about the cost of having two employees who do nothing but update the deduction records of the firm's 1,375 employees. These two employees earn an average of $24,300 each. The controller decides to try having the employees enter their own deduction changes. To do this, an interface must be written to the payroll software that allows access by all employees with an employee number and a password. The programming cost of this interface will be $35,000. The interface will allow any employee with computer access to change her deduction records. Unfortunately, 90% of the staff comprises production workers with no access to a computer. Accordingly, the controller determines that computer kiosks may be constructed at various locations in the facility for access by the production staff. These kiosks will allow access by all staff and will cost $5,000 each to install. Four kiosks will be needed. General training sessions will be held for the staff during their lunch breaks. The cost of the trainers and training materials will be $3,500. Should this project be initiated?

Solution.   The cost of the project is related to creating an interface for employees to use, plus installing computer kiosks and paying for trainers to educate the staff in the use of the equipment. The savings are entirely from eliminating data entry payroll costs.

Cost of Installing Deduction Interface

Cost/computer kiosk	$ 5,000
No. of kiosks	× 4
Total cost of kiosks	$20,000
Programming cost	$35,000
Training cost	$ 3,500
	$38,500
Total cost of deduction interface	$58,500

Benefit of Installing Deduction Interface

Cost/yr of data entry clerk	$24,300
Clerical positions eliminated	× 2
Total savings from deduction interface	$48,600

With costs of $58,500 and annual savings of $48,600, this project has a payback of 1.2 years and should be implemented.[2]

Install Bar-Coded Time Card System

Katherine Peterson, general manager of Sucker Candy Corp., notices that the company employs an accounting clerk to do nothing but enter time card information for the firm's 347 production workers, who operate the candy manufacturing equipment during all three shifts. This clerk also corrects mistakes made during the original data entry. Investigation with the various manufacturing supervisors reveals that one time card in eight is incorrectly entered by the clerk, resulting in complaints by the manufacturing staff. The controller contacts a manufacturer of bar-coded time card scanning equipment and finds that making a bar-coded, laminated time card for each employee will cost $1.50. The firm will also need a pair of bar-code scanners, to be located next to each exit door. The scanners cost $1,800 each. The scanners will have to be linked to a dedicated personal computer, which collects the information. The PC, including network hookup charges, costs $3,000. Cabling must also be purchased and installed to link the two scanners to the PC. The total cabling costs are $2,800. Finally, the controller must purchase interface software for $2,000 that links the payroll information on the PC to the payroll software that prints paychecks. If this system is installed, the time card entry clerk will be reduced to half-time and will be assigned the task of reviewing time card entries on the time card PC for errors. The time card entry clerk earns $18,650 per year. Should the bar-coded time card system be installed?

Solution. The cost of the system is related to purchasing the time cards, scanners, cabling, PC, and software interface. These are all one-time costs, except for purchasing additional bar-coded time cards for new employees. The continuing savings are earned by reducing the amount of data entry payroll.

---

[2] The largest variable is the training cost. This example assumes that training occurs during lunch breaks. If not, then paying the staff to attend training is a much more significant cost. Also, periodic retraining may be necessary, which also adds to the cost.

Cost of Installing Time Card System

No. of production employees	347
Cost/time card	× $1.50
Total cost of time cards	$ 521
Cost/bar-code scanner	$ 1,800
No. of scanners needed	× 2
Total cost of scanners	$ 3,600
Cost of personal computer	$ 3,000
Cabling costs	$ 2,800
Cost of software interface	$ 2,000
Total time card–related costs	$11,921

Benefit of Installing the Time Card System

Cost/yr of data entry clerk	$18,650
Percentage of year no longer needed	× .50
Total labor savings	$ 9,325

With costs of $11,921 and annuals savings of $9,325, the bar-coded time card system will have a payback in 1.3 years. This project should be accepted.[3]

Deposit Paychecks Directly to Employee Bank Accounts

Mr. Baumgartner, president of Sales on Call, a telephone wholesaler, receives many complaints from his sales staff, which is constantly traveling and therefore unable to pick up paychecks at the office on payday. Mailing the checks is not a viable alternative, for many salespeople are out of town so much that they cannot check their mail on a timely basis. Currently, many salespeople cannot pay their credit card bills on time because they are not in town to cash their paychecks, resulting in a continuing credit problem while traveling on company business. To solve the problem, Mr. Baumgartner is considering using direct deposit to electronically wire money directly into the sales staff's bank accounts. The company employs 42 salespeople. The cost to make a direct deposit is $0.75 per paycheck deposited. The company pays employees every other week. In addition, Mr. Baumgartner would like to wire payments to salespeople for their weekly expense reports. The entire sales staff travels every week, excluding

---

[3] The primary variable in these costs is the existence of interface software to the company's payroll software. If that does not exist, time card information will either have to be manually transferred to the payroll software (which involves extra data entry cost) or a custom interface will have to be created (which involves considerable programming costs). If either cost is incurred, the payback period will lengthen considerably.

their two-week vacations. If the direct deposit system is implemented, Mr. Baumgartner expects to stop issuing an average of $1,000 in advances to the sales staff to cover their expenses because they have not received their paychecks. On average, 25 salespeople are in receipt of these advances at all times. The company borrows money at a 9.5% interest rate. Should Sales on Call implement a direct deposit system?

Solution. The cost of direct deposit is the transaction fee charged by the bank and multiplied by the number of direct deposits per year. The savings are from the interest expense saved on advances made to salespeople to cover their lack of cash when they cannot pick up their paychecks.

Cost of Direct Deposits

No. of salespeople	42
No. of pay periods/yr	× 26
No. of pay deposits	1,092
No. of salespeople	42
No. of expense reports/yr	× 50
No. of expense deposits	2,100
Total no. of deposits	3,192
Cost/direct deposit	× $.75
Total cost for direct deposits	$2,394

Benefit of Direct Deposits

No. of salespeople with advances	25
Outstanding advance/yr	× $1,000
Advances outstanding	$25,000
Interest cost	× 9.5%
Interest cost saved	$ 2,375

In short, with costs of $2,394 and savings of $2,375, there is no clear savings from implementing a direct deposit system. This is a common issue with direct deposit, especially when it is implemented for all corporate staff rather than just the salespeople. It can be presented as an employee benefit, or the advantages of avoiding a hassle with getting paychecks to traveling employees can be presented as an unquantifiable benefit.

Remit Paychecks to Employees by Electronic Mail

The controller of the EverKlear Windshield Company notices that the cost of mailing direct deposit remittances to all company employees is 80 cents per paycheck, including postage costs. The company has 650 employees and a one-week pay cycle. The controller notices that half of the company's employees have access to electronic mail. By sending direct deposit remittances by e-mail, the company can avoid the mailing costs associated with sending the information on a paper form. The programming staff reviews the situation and calculates that it will cost $20,000 of programming time to create a batch file that extracts payroll information from the payroll database and sends individual records to employees through electronic mail. Should this project be completed?

Solution.   The cost of the project is entirely related to programming fees, while the savings are from the labor, envelopes and paper, and postage needed to mail out remittances.

Cost of Remitting Paychecks by Electronic Mail	
Programming cost	$20,000

Benefit of Remitting Paychecks by Electronic Mail	
No. of employees	650
Percentage of employees with e-mail	× .50
No. of employees with e-mail	325
Mailing cost/employee/remittance	× $.80
Total cost/pay cycle	$ 260
No. of pay cycles/yr	× 52
Total cost of remittance mailings	$13,520

With costs of $20,000 and savings of $13,520, this project has a payback of 1.5 years and should be completed.[4]

Include Vacation and Sick Time Accruals on Payroll Remittances

Ms. Athelwaite, supervisor of the payroll department of Monkey Around Lubrication Shops, is tired of having the company's hundreds of mechanics call or visit the payroll department during their break times to inquire about available sick and vacation time. She estimates that the average employee inquires about this

---

[4] The key variable is the percentage of employees with access to electronic mail. The payback period tends to be shorter in service industries such as insurance, where everyone has access to a computer. The payback period is longer in the manufacturing industry, where people do not have such access.

matter for three minutes once every other month. The company has 600 employees. The average payroll clerk who must answer the vacation and sick time questions receives a burdened pay rate of $15.42 per hour. To fix the problem, Ms. Athelwaite talks to the company's programming department and finds that a remaining vacation and sick time accrual can easily be added to everyone's payroll remittance advices (which are received weekly) for 30 hours of programming and testing time by a programmer. The programmer earns a burdened rate of $28.75 per hour. Is it worthwhile to implement the programming project?

Solution. The total annual cost is the time spent by the payroll clerk to research the time accrual for each employee who inquires. It is assumed that employees are inquiring during their break periods, so that this does not interfere with their work (which would otherwise be an additional cost). The only extra cost is the one-time programming cost to add the vacation and sick time accrual fields to the remittance advice.

Cost of Remittance Change

Time to program change	30 hrs
Programming cost/hour	× $28.75
Total cost of programming change	$862.50

Benefit of Remittance Change

No. of employees	600
No. of inquiries/employee/yr	× 6
Total no. of inquiries/yr	3,600
Time/inquiry	× .05 hr
Time/yr for all inquiries	180 hrs
Cost/hour of payroll clerk	× $15.42
Total cost for inquiries	$2,775.60

Thus, the total cost is $862.50 and the total savings are $2,775.60, resulting in a payback of the programming investment in one-third of a year. Also, the cost will not be incurred again, while the savings will continue to accrue. Based on the cost/benefit analysis, the programming change should be implemented.

In summary, improvements to the payroll system involve the purchase of additional computer equipment or the expenditure of fees for programming. These costs can be justified in most cases, depending on such variables as the access of staff to computers, employee possession of bank accounts, the amount of training needed, and the percentage of the staff that is paid from a time card. The most difficult project to justify based on costs alone is the use of direct deposit to pay employees.

REPORTS

The payroll department is usually awash in reports, so it is not a service to that department to come up with even more reports. However, there are two simple reports that contribute to the efficiency of the department.

The first report is a *list of employees who are not using direct deposit.* Since using direct deposit reduces the work load of the payroll department, the payroll manager should print this list of nonusers regularly and contact those employees to see if they can be persuaded to change to receiving payments directly into their bank accounts.

The other report is *a list of manual changes to payroll records being made by the payroll department* (see Table 6.3). Since the intent of many of the changes in this chapter is to eliminate the data entry tasks of the payroll department through bar coding, automation, and having employees make their own deduction changes, it is important to determine what kinds of changes are still being made by the payroll staff, so that those changes can be targeted for elimination. This report requires a program to collect information about which employee records are being accessed by terminals in the payroll department.

In short, very few additional reports are needed by the payroll department, but two reports can be added that tell the payroll manager how to further improve the processing speed of the department by increasing the number of employees who receive direct deposit and by reducing the number of manual changes to employee records by the payroll staff.

TABLE 6.3   Payroll Changes Made by the Payroll Department

Terminal No.	Date	Employee No. Accessed	Employee Name Accessed	Record Accessed
00140	05/05/97	001872	Smith, Barney	Payroll deduction
00140	05/05/97	003021	Davis, Freeman	Payroll hours
00140	05/06/97	001101	Gregory, Melvin	Benefit deduction
00140	05/06/97	009234	Orroway, Adelaide	Payroll hours
00145	05/05/97	001103	Brando, Elvis	Payroll hours
00145	05/05/97	001103	Smith, Keith	Benefit—health
00145	05/06/97	003378	Danbury, Irwin	Benefit—health
00150	05/06/97	009821	Rae, Norma	Benefit—dental

TECHNOLOGY ISSUES

Technology plays a part in most of the changes advocated in this chapter. However, some (such as employee kiosks for deduction changes) only require the addition of more computer terminals to the company, while others (such as automating commission calculations) merely involve creating computer programs to replace previously manual calculations. The only truly technological breakthrough that has a significant effect on the payroll function is the use of bar coding or magnetic stripes to track employee time. Bar coding and magnetic stripes have been used by a number of industries for several decades, so they cannot be considered new. Nonetheless, many companies have never used them, so a brief description of the process is included here.

Bar coding converts a set of alphanumeric digits into a band of printed bars that uniquely identify the digits with bars of varying thickness and separation. When scanned with a laser, this bar code is converted back into alphanumeric digits. Since most scanning equipment is attached directly to a computer terminal, the computer sees the incoming information as having been entered through the keyboard, so the computer is indifferent to the presence of the bar-coding equipment. It is nearly impossible for the bar-code scanning equipment to incorrectly translate a bar code into the wrong alphanumeric information, which gives the system a high degree of reliability. There are just two problems with the system: first, it is possible to create a bar code with the wrong information, so that the equipment correctly translates the incorrect information and feeds it to the computer, and second, a poor-quality bar code cannot be scanned.

Magnetic stripe technology is the same in principle as bar coding. A set of alphanumeric digits are encoded on a magnetic stripe, which is converted back to the original text when run through a scanner. The primary difference between the two types of technology is that a company can easily create its own bar codes on a laser printer, whereas magnetic stripes must be manufactured for the company by an outside supplier. This makes bar codes much more flexible for internal use.

Bar codes and magnetic stripes are most commonly used for applications where exactly the same information is entered into the computer over and over again, so that keypunching errors can be avoided by simply scanning the information from a bar code or a stripe that already contains the information. One of the best applications is time collection, where the same information (an employee number) is scanned into a computer that adds the current time. A bar-coded job number or activity code from a standard list can also be added to the entry, resulting in an error-free transaction that requires no use of a keyboard at all.

In short, bar codes or magnetic stripes can be used to send error-free information to the payroll department's time-tracking system with no errors and no keypunching by the payroll department.

MEASURING THE SYSTEM

The following measurements are useful for tracking the cost and payment speed of the payroll system.

Improving the Amount of Data Input

If the quantity of information input is reduced, the cost of inputting it should also go down. Therefore, tracking the total payroll department cost is a good measure of the efficiency of the department. However, a better way to create a baseline of how much the department costs are changing over time is to track payroll costs per billions of revenue dollars, or payroll costs per employee. The primary payroll cost should be the pay of the payroll clerks and related benefits, though charges for office supplies, office space, and computer usage may be added as well. A corporate overhead charge should not be used for this calculation, since the amount of that charge is not controllable by the payroll department.

Another way to measure the amount of data input is to track the total number of manual payroll changes of any type made by the payroll department as a whole by month. This information is most easily collected by installing a program that determines the number of changes made by computer terminals located in the payroll department. This information tells the payroll manager how many changes are being made to records even though employees may be allowed to make changes themselves. A detailed listing of which records were altered can then be reviewed to determine what types of changes are still being processed by the payroll department. These changes can be investigated and action taken to reduce those types of changes in the future.

Improving the Speed of Payment

Switching employees to direct deposit reduces the check signing and paper cost of the company. The easiest way to track the company's progress in converting to direct deposit is to add up the number of direct deposits per pay cycle and divide that amount by the total number of payments, yielding the percentage of employees receiving direct deposits.

In summary, a few simple measurements are needed to provide an effective set of statistics about the performance of the payroll department. Collecting information for these measurements is not difficult (with the possible exception of the number of payroll changes) and provides management with enough information to spot departmental efficiency problems very quickly.

IMPLEMENTATION CONSIDERATIONS

A number of issues related to change management may occur when the revised payroll system is installed (see Chapter 11). The following problems can be expected when implementing a streamlined payroll system.

Give Employees Direct Access to Their Deduction Information

Programming an interface to the deductions database for employees to use, even one that operates over a telephone, is by no means impossible. The difficult part is making people use it. Of course, the company can create a policy that forces employees to only use direct access, but then they may go out of their way not to access the system, resulting in deductions that employees are not happy with. This reluctance is caused by a very large number of employees still not feeling comfortable with using computers to access their deduction information (how many people still do not use ATMs?). One solution is to link the direct access system to a touch-tone telephone system, which employees may be more comfortable with than a computer. Another method is to use training that incorporates having people access their records through the direct access system. Having them train in a hands-on mode gives them more comfort with the equipment.

The interface used by employees to access their deposit information should be thoroughly tested prior to letting employees use it; otherwise, the system may show employees the deduction information of other employees or fail to update their records with input changes.

Adopt an Automated Commission Structure

The biggest problem with automating the commission system is that the programming team will finish the programming work, have the sales manager review the first working model, and then discover that there are one or two special situations that simply have got to be added to the calculations. This is also known as "scope creep," whereby the initial system is expanded to include requirements that were not built into the original project scope. This is a common problem with commission systems, because the sales management group tries to continually add variations that will motivate the sales team to ever-greater performance. In fact, it can be expected that "essential" changes will still be lobbied for even after the project has been officially completed. The best way to avoid this implementation problem is to have the sales manager sign a list of program features when the program's scope is first being determined, and hold the project to the features list despite all distractions. The best way to avoid subsequent changes is to have all corporate programming requests submitted to a programming prioritization committee that determines the value of all submitted requests in relation to the pool of other requests. With this system, low-end requests never percolate up to the top of the list.

Install a Bar-Coded Time Card System

A bar-coded time-keeping system is quite a shock to a production force that has been used to punching a time card. The biggest problem is lack of a visible record that shows the employee that the time is actually being recorded. A good way to gain the cooperation of the production group is to involve a small group of production employees in a test of the system, whereby they run a badge through a scanner and then go to a PC terminal and see their time being recorded. This small group can then act as opinion leaders in spreading word among the production staff about how the system works. Any problem in recording time will be a devastating setback to acceptance of the system, since the company has just failed to record the single most important thing in the work lives of the production staff—the time for which they are paid. To avoid this problem, make sure that the system is thoroughly tested before switching systems; perhaps ask the production staff to use both the old and new systems at the same time until the new system is "bug-free"; and always make certain that there are power backup systems that will keep the time card system running even when the company's main power systems are shut down, so that data cannot be lost. In addition, there should be daily backups of the payroll database, so that a loss of information will require the minimum amount of data reconstruction.

Another consideration when implementing a time card scanning system is the number of scanning workstations. If just one station is used, there may be a very long queue of workers lined up to clock in or out (though scans with a bar code are faster than punching into a clock). The alternative is to set up scanning terminals in multiple locations throughout the plant. If multiple stations are used, the controller must be sure to budget for stringing a considerable amount of extra wiring overhead so that the terminal is linked to the controlling computer. Also, if the production staff is expected to scan themselves in or out of a job, a bar code sheet should be provided next to each scanning station that lists all current jobs. Then a worker scans the time card to clock out of the last job, scans it again to clock into the next job, and scans the bar code of the next job with a portable scanner to identify the next job. It is important that there be a sufficient number of scanning terminals if employees have to perform multiple scans during the day to identify their time on various types of work—otherwise, they will spend a startling amount of time walking to and from the scanning terminal, or they may not bother to go to the terminal, resulting in all their time during the day being charged to one activity instead of to a variety of activities.

Combined Databases

When merging the databases related to human resources, benefits, and payroll into one database, it is important to create backups of all the old databases as well as of reports showing all the information in those databases. This information is needed when cross-checking the information in the newly created database for errors. There can be errors related to keypunching, records entered more than once, and (most devastating for the affected employees), no record at all. Each of these types of errors can have a significant effect on an employee, since the accounting staff may have to request new forms to be filled out by the employee to create a new set of information; while this missing or altered information went undiscovered, the employee may have lost benefits.

Send Payroll Payments by Direct Deposit

Direct deposit transfers do not always work on the first try. This is rarely caused by an error in transmission by the bank. Instead, it may have to do with an account not being opened yet, the bank code being incorrectly set up, or the list of payments not being sent to the bank in time for any new account transfers to be set up. Once bank identification numbers and account numbers are set up for new employees, it is rare for payments not to go through to bank accounts in a very reliable manner—but that first time can be difficult. To cover all eventualities, the new employee should be warned that the direct deposit may not go through on payday. The company will be told by the bank if no deposit was made, so the company can issue a manual check to the employee a day late to cover the problem. The employee should be warned about this backup system in advance to avoid concern by the employee that payment will not be made at all.

In summary, the biggest problems with payroll-related changes involve proper testing of the changes and training employees thoroughly about the new changes. These two issues are particularly important in the area of payroll, for this is the system that all employees are affected by on payday. If the streamlining changes do not work, then the failure becomes obvious to all employees right away, and a failure in this area can lead to intense employee dissatisfaction.

SUMMARY

This chapter concentrated on improving the efficiency of the data entry portion of payroll, since there are many detailed manual calculations and keypunching required at that stage. By implementing employee-controlled deduction entries, bar-coded time keeping, simpler commission calculations, and combining databases, it was demonstrated that the data input phase of the payroll function could be speeded up. Improvements later in the process include merging databases, including vacation and sick time accruals on the payroll remittance, and sending remittance information by electronic mail. All these improvements reduce the work load of the accounting department, reduce the error rate, and contribute to the speed of payroll processing.

REFERENCES

Hackett Group. *AICPA/THG Benchmark Study: Results Update and Analysis.* 1994.

"The New Blueprint for Finance." *CFO Magazine,* June 1993, p. 32.

7

THE BUDGET

The process of creating the annual budget consumes a large amount of managerial time, not only to determine budget assumptions but also to decide on sales projections, the need for additional facilities, financing needs, and staffing concerns, and to critique the plans of other departments. This series of meetings may be repeated at various times during the year if revenues depart so far from the budget that the associated expense budgets are no longer relevant. Further, other meetings may take place that focus on how actual revenues and expenses are turning out in comparison to the budget. In total, the time allocated to the budget process is among the largest draws upon the time of today's manager. Reducing that block of time will allow the manager more time to run the business. This chapter discusses how to reduce the time needed to create, modify, and monitor the budget.

CURRENT SYSTEM

This section gives an overview of the budget process. It shows how the various types of budgets are linked, and how a change in one budget causes changes in other budgets. Also, it notes the inefficiencies that are built into the typical budget process.

The budget cycle begins with a management meeting that reviews the firm's strategic objectives. The strategic objectives must be agreed upon first, since they may involve such issues as eliminating entire businesses or creating new ones, which have obvious ramifications on the budget. The strategic objective meeting then leads into a (usually prolonged) discussion of revenue objectives for the upcoming year. This is an important part of the budget, for all expenses are derived based on revenue; this is because the revenue number tells the budgeting team how much money can be spent to ensure that the revenue goal is met and that there will be a profit. The revenue goal also requires a detailed analysis of sales targets for sales regions and salespeople, which may entail a change to the commission structure and a change in the number of salespeople, the method of selling, and an expansion into entirely new sales territories (e.g., overseas sales).

With a preliminary sales plan completed, the budget team changes its composition to include the production staff and works on a production plan. The production plan determines how many materials must be purchased (taking into account the amount of inventory currently in stock), how many workers must be employed, and if additional facilities must be constructed in order to create the number of units required by the sales plan. In many cases, the completed production plan is the first point in the budget process when the management team learns what the cost of goods sold will be, which may send the budget back to the managers for a second discussion of the amount of revenue to be budgeted (and may send them all the way back to the strategy level to make changes at that point). If the revenue plan changes, then the production plan must be changed as well.

Once there is general agreement on the production plan, it is handed to the other company departments, such as engineering, administration, and management information systems. These departments construct budgets based on the revenue and production plans. These plans include expenses (typically with a heavy emphasis on personnel costs) as well as fixed asset purchases. When combined with facilities requirements that are included in the production plan, the budget team has enough information to construct a capital expenditures budget. This concludes the primary set of budgets, with one exception.

The final budget is the financing plan. This budget takes the cash flow information from all previous budgets and the cash requirements of the capital expenditures plan, and derives working capital needs based on days of receivables and payables as well as inventory turnover. With these inputs, the financing plan comes up with either the amount of cash spun off by the company or the amount required. This information is reported back to the management team, which frequently alters the budget to take into account the realities of financing. If cash requirements are too high, then some high-cost projects may be eliminated. If the company's basic cash flow is too small, then expenses may be slashed at the departmental level or prices may be raised. No matter what changes are made, it is likely that management will review and re-review the budget once the full set of plans are developed, until an acceptable balance of funding needs and profitability is reached. A flowchart of the usual set of plans that make up the company budget is shown in Figure 7.1.

In summary, the typical budget process requires a large number of iterations and many meetings by the management team before a satisfactory budget is created. This is not an efficient use of managers' time, since they must meet many times to make incremental changes to the budget and than wait for the budget

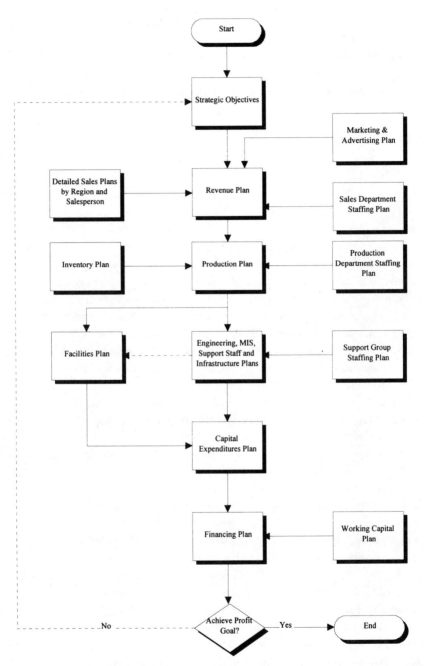

FIGURE 7.1 Typical budget process.

team to process the changes and give the results back to them. This is also not an efficient use of the time of the budget support staff, since they must make all the detailed changes to the plan once incremental changes have been approved by management. The following section discusses ways to reduce the number of budgeting iterations as well as ways to reduce the work load of the budget support staff.

REVISED SYSTEM

This section focuses on streamlining the budget process. This can be done by carefully determining company strategy and prioritizing related projects *before* detailed expense budgets are created, by carefully outlining the timing of the budget process in a budget manual, implementing flexible budgeting, and finding ways to make the budgeting of expenses more accurate so that the budget does not have to be recast partway through the year. When implemented, these changes will allow the company to prepare a detailed budget with fewer iterations.

When beginning the budget process, the company should immediately *limit the budget with the maximum amount of funding that will be available in the upcoming year.* This is necessary to keep management from spending considerable time devising a strategy that will require immense capital resources that cannot be obtained. As long as the management team is aware of the amount of cash that will be available, it will reduce the range of possible budget activity levels and more quickly arrive at an appropriate budget for the upcoming year.

Detailed expense items should not be budgeted until the high-level budget is approved. When a detailed expense budget is brought by each department head to the initial companywide budget meeting, this is putting the cart before the horse, for the departmental budget cannot be created until the corporate revenue figure has been determined. Once revenue, which drives the level of corporate activity, has been decided upon, it is a straightforward matter to put together a departmental budget. However, even when revenue has been agreed upon, it is likely that the revenue amount will be altered as the management team goes through a number of iterations with the budget model. Given the number of iterations, the number of subsidiary expense budgets may also be recast many times. To prevent the departments from wasting too much time redoing budgets, it is best to create a summary budget and then issue a detailed budget only when management has signed off on a final overall corporate budget. It is easy for many departments to create and maintain a summary budget merely by tracking the number of employees in the department; the largest cost of most departments relates to salaries and benefits, with other costs such as office supplies varying in direct proportion to the number of people in the department. Therefore, by rebudgeting based on the number of people in the department (a quick and easy exercise), the department head can rapidly recast a reliable department budget without having to conduct a line-by-line review of all expenses in the budget. A departmental budget based on headcount is shown in Table 7.1.

The high-level budget should include an analysis of working capital. Too many budgets do not include the cash flow requirements of working capital. When it is included, management suddenly realizes that the strong growth plans of the up-coming year cannot be realized for lack of capital and the budget must be recast once again to find an activity level that is supportable with less cash. Working capital can be determined even without a complete balance sheet. A simple list-ing of receivables, payables, and inventory is sufficient for a quick working capi-tal analysis. The working capital portion of a high-level budget can be as simple as the example shown in Table 7.2.

TABLE 7.1 Engineering Department Budget Based on Headcount

Job Title	No. of Employees	Average Salary	Extended Salary	Benefits Percentage	Total Employee Cost
VP Engineering	1	$125,000	$ 125,000	20%	$ 150,000
Engineering Super.	3	75,000	225,000	20	270,000
Engineer III	15	60,000	900,000	20	1,080,000
Engineer I	15	45,000	675,000	20	810,000
Designer	12	35,000	420,000	20	504,000
Librarian	2	30,000	60,000	20	72,000
Secretary	3	25,000	75,000	20	90,000
Total staff	51		$2,480,000		$2,976,000

Variable Expense Items	Cost per Employee	No. of Employees	Total Variable Cost
Office supplies	$ 450	51	$ 22,950
Travel	2,570	51	131,070
Telephones	350	51	17,850
Total variable			$171,870

Fixed Expense Items	Total Fixed Cost
Subscriptions	$ 500
Dues	1,000
Annual awards	350
Total fixed	$1,850

TABLE 7.2　High-Level Working Capital Analysis

Account Type	This Year Turnover	Next Year Turnover	This Year Working Capital	Next Year Working Capital
Revenue			$30,000,000	$50,000,000
Cost of goods sold			9,000,000	15,000,000
Receivables	45 days	40 days	3,750,000	5,555,000
Inventory	10 turns	12 turns	900,000	1,250,000
Payables	35 days	35 days	875,000	1,458,000
Total working capital			$ 3,775,000	$ 5,347,000

Most budgets contain a significant number of projects. If funding in the budget is not sufficient to allow all projects to be completed, there is usually a great deal of time-consuming haggling among managers to determine which projects are to be funded and which are to be moved off to a future year. To avoid this problem, one of the first tasks of the management group after strategic goals have been set is *determining how well projects match the company's strategic goals.* If there is no close linkage between strategy and goals, then the availability of cash has nothing to do with a project being completed—it should be eliminated just because it does not support the company's strategic direction. By reviewing projects as soon as possible in the budget process, a number of projects can be removed from the budget without any additional need for detailed expense justifications, thereby decreasing the work of the budget analysts.

Another item that should be derived early on in the budget process is *the nature and amount of underlying costs.* An underlying cost is any cost that is likely to exist during the upcoming year unless management takes specific action to remove or add to the cost. Examples of underlying costs are utilities, rent or lease payments, depreciation on existing assets for the upcoming year, taxes, and benefits costs for employees. This information can be prepared far in advance of the rest of the budget information to save time and can be treated as the underlying amount of cost that must be covered by the gross margin. Management can also review the information early in the budget cycle and decide if alterations must be made. For example, if depreciation is too high, then management can decide to sell some assets. If benefit costs per person look excessive, then the human resources department can begin to review these costs well before the rest of the budget work begins.

Having identified projects that specifically relate to corporate strategy, it is now necessary to *break out those costs that must exist, irrespective of all but the most significant strategy shifts.* This can constitute a significant portion of all expenses, and must be identified so that management can determine the fixed costs that must be covered. These costs include long-term leases, long-term contracts to purchase raw materials, and long-term rental agreements. They may not include any employees or short-term costs if part of the strategic vision is to sell off entire businesses. However, since the sale of businesses would have already been determined as part of the strategic plan, it would be apparent if this were to be the case. In the absence of a corporate sale, it would be acceptable to list the payroll and related costs of maintaining the company's operations at any activity level. This information must be determined as soon as possible, so that management can concentrate on budgeting for any remaining costs that vary with the level of corporate activity.

The budget process should be organized as much as possible with a *budget manual.* This manual, distributed at the initiation of the budget cycle, should include due dates for all portions of the budget, who is responsible for each part of the budget, and what line items in each part of the budget must be completed. A sample budget should also be included, as well as sample budget forms and a calendar that lists all due dates. This degree of organization is particularly necessary for companies with subsidiaries, where there are many people in outlying areas who must send in their budget information before the overall budget can be created. Even in a smaller company where the budget analyst is within reach of all departments, it is still a good idea to introduce some rigor to the process, so that there can be no excuses by the staff regarding the lateness of budget information.

It is usually possible to *reduce the number of meetings.* Budgeting involves lots of meetings to go over changes that are really the responsibility of just one department. That one department is quite capable of making revisions related to its own costs (e.g., the production, sales, or marketing departments) without input from other parties who have far less knowledge of the department. Within the department, it is also not necessary to have an excessive number of meetings. Instead, the head of the department should assign an expert to the task of making budget revisions for the department. This person (frequently the department head) is the one most capable of identifying budget costing errors or estimating changes based on alterations to the level of corporate activity. Thus, it is possible to avoid meetings at both the interdepartmental and intradepartmental levels, thereby saving valuable management time.

Having advocated a reduced number of meetings, it is also critical to *have a well-represented meeting that reviews strategies and assumptions.* This meeting is needed to bring together experienced representatives from all major departments to critique the major assumptions of the budget for the year before any detailed work is performed at all. This meeting keeps the budget staff from spending too much time creating detailed plans based on false assumptions. By ensuring that the correct high-level direction is being taken right from the start, less time will be wasted in constructing the budget.

Another change to consider is *reducing the budget from a monthly to a quarterly plan.* Top-level management budget meetings can frequently degenerate into an argument about the specific months in which revenues and related expenses will occur. If the number of budget periods are cut to one-quarter of the previous number, the amount of discussion related to the seasonality of revenues and expenses can also be reduced.

A good way to reduce the work of creating the budget is *reducing the number of accounts.* This can be pursued throughout the year rather than just during the budget process. The intent of having fewer accounts is that there are fewer line items in the budget that must be estimated, fewer line items that can be incorrectly entered, and fewer line items that can be incorrectly summarized by bugs in the budget model. In short, fewer accounts shrink the work of preparing the budget.

A tremendous time-saver is to avoid a fixed budget and instead *use a flexible budget.* A fixed budget derives a profit amount based on a very specific revenue level for the year. Expenses are tied to that revenue level. If the revenue actually earned varies appreciably from the amount budgeted, the entire budget must be recast to see what effect the revised revenue will have on related expenses. Every time the budget is recast, valuable management time is lost in reviewing the budget, debating changes, and confirming the new budget. With a flexible budget, it is possible to designate both fixed and variable portions of some expenses and have the variable portions change as revenue changes. This allows the company to quickly change to a new budget with minimal work by management. It must be noted, however, that flexible budgeting only works within a relatively narrow range of revenue figures—once revenue departs significantly from the expected target, there may be a significant change in the related expenses in terms of headcount and facilities that will require a complete recasting of the budget.

Determining cost drivers allows the company to develop a budget that reacts well to changes in activity levels, which is needed in a flexible budget. A cost driver is an activity that changes a cost. If the company finds, by constantly comparing budgeted costs to actual costs, that specific activities dramatically change costs, then those drivers should be budgeted. For example, the number of employees on-site may have a direct effect on utility costs. Therefore, the budget should link the number of on-site employees to changes in the utilities expense. Similarly, the number of employees can be linked to the cost of office supplies. This cost driver yields a more accurate cost than linking either utilities or office supplies to changes in revenue. By using cost drivers, the budget becomes more refined and requires fewer laborious changes and variance analyses over its life.

Cost drivers can be taken to greater extremes in the budget process by *using activity-based costing* (ABC). Though rarely used in budgeting, it can provide valuable refinement to the process by linking expense changes to variations in the product mix, geographical sales mix, and customer mix as well as other permutations. However, the price of increased accuracy is a more complex budget model; the controller should be aware that increased budgetary complexity may lead to errors in the model that require considerable manual recalculation to discover. Another problem is that off-the-shelf budgeting packages do not normally include activity-based costing features. Thus, the controller who uses ABC may be forced to develop a custom-designed budget program, which requires time for debugging.

Once an acceptable budget has been agreed on, the budget team has one remaining task. It must ensure that *performance measurement and reward systems are in place.* This is necessary to ensure that the company's performance against the budget is tracked and fed back to management, and that certain employees are rewarded for their ability to meet budget objectives. Without a performance tracking and reward system, the company will be more likely to not achieve budget goals, necessitating a recasting of the budget later in the year to bring expenses into line with a reduced level of corporate activity. Since the budget team wants to avoid a lengthy recasting of the budget, completing the performance tracking and compensation system should be considered a high priority.

In summary, the most time-consuming part of the budget process is the number of iterations required to arrive at a budget model that is workable in light of funding needs and corporate resources. This section reviews a number of ways to create a good budget the first time, thereby eliminating a large part of the expense and wasted time associated with creating several versions of the budget. It also shows how to derive more accurate expenses for the budget, so that expenses do not have to be recast at a later time during the year. By doing it right the first time, the cost of creating a budget can be significantly reduced. A streamlined budget process is shown in Figure 7.2.

CONTROL ISSUES

Unlike many of the other accounting-related areas covered in this book, the budget cycle has no transactions. Since there are no transactions, there is no need for controls in the budget area.

However, it is important once the budget has been finalized that expenditures in excess of the budgeted amount are highlighted for the benefit of management. There are several ways to accomplish this. First, the company's general ledger software should include a feature that lists the budget amount for each account next to the actual expense for the month and the year-to-date. Most software packages contain this reporting feature, but some controllers do not use it. The report is sent to department supervisors, who compare budgeted to actual costs.

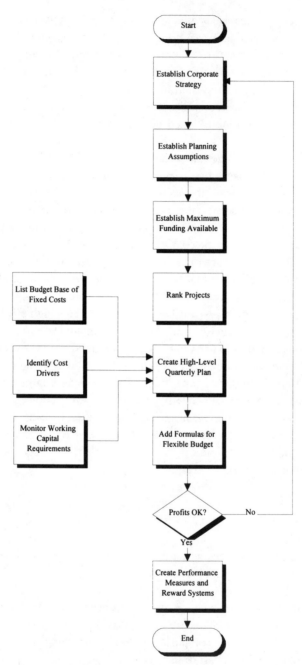

FIGURE 7.2   Streamlined budget process.

The second way to track expenditures in excess of the budget is to have a drill-down feature in the accounting software. A drill-down feature presents the user with the detailed costs that make up a summary cost balance, usually by jumping directly from the summary screen to a detailed expenditure screen. This concept can be applied to a report that lists the net change in an account during a month and details of the exact expenses incurred during that period. A third technique is to send a report to management that lists all expenditures over a certain dollar amount during the period, with a detailed description of the expenditure. A typical minimum dollar amount for such a report is $1,000.

Also, once the budget has been finalized, it is important to institute controls over major expenditures that were noted in the budget. Even though the budget has already been approved by management, individual large purchases or pay increases should still be justified to management in detail and not implemented without the approval of management. This is because the intent of the type of budgeting presented in this chapter is to create a budget without getting into extreme detail, thereby allowing the management team to complete a quality budget very quickly. If capital expenditures and pay increases that were listed in the budget at a macro level (e.g., "unspecified production equipment for $100,000," or "accounting department pay increase of 5%") are implemented without further controls, then this defeats the purpose of the quick budget, which is to decide on an appropriate level of corporate activity for the upcoming year, not to act as an expenditure approval document. The way to control this problem is to require management approval of individual asset purchases and pay increases.

In short, no controls over the budget process are needed, but it is important to track expenditures as they compare to budgeted expenses and to institute tight management review controls over spending amounts that were approved *in concept* during the budget formulation process.

QUALITY ISSUES

In this book, good quality is defined as lack of errors. There are a number of situations in which errors can creep into the budget process, yielding inaccurate information. This section reviews common errors in the budget process and how to avoid them.

Merging Budget Spreadsheets

One common source of errors is merging budget spreadsheets that have been sent in from various subsidiaries. This problem occurs when the person in charge of budgeting sends a standard electronic budget form to subsidiary locations, with instructions to fill in the blanks that have been prepositioned on the form. The subsidiaries frequently have special situations that call for an additional line item. They include the exception line items and send the electronic spreadsheet back to the budget analyst. When the budget analyst merges the documents, there is now an additional line item that increases the length of the spreadsheet by one line, creating incorrect line and row totals and generally requiring considerable work to fix. There are several ways to reduce the error rate resulting from this problem.

Reduce the Number of Accounts.   By shrinking the number of accounts in all subsidiaries, it becomes very unlikely that additional line items will be added to the budget. The corporate parent can even issue a standard list of accounts that must be used by all subsidiaries. The problem with this approach, however, is that subsidiaries in different industries may have operations so radically different from each other that a single chart of accounts for the entire company is not possible without having an extremely lengthy list of accounts.

Enter Budget Information Directly into a Central Database.   The company can create a central database file and allow access to it from all subsidiary locations, so that information does not have to be transferred from paper or electronic documents at all—the staff of the subsidiary does all the data entry. The problem with this approach is that a centralized account code structure must be imposed upon all subsidiaries, and a budget database must be created at corporate headquarters, which can be quite expensive.

Merge Only Summary Information.   A problem with merging all the subsidiary budgets is that, from a management perspective, the corporate staff is most interested in summary-level information and therefore does not care if it receives detailed line item information. Most headquarters-level analysts restrict their reviews of subsidiary budgets to the subtotals and totals of the various budgets. These tend to be the same across all industries (e.g., total headcount, total marketing, and total sales costs are summarized in most financial statements) and are therefore much easier to summarize at the headquarters level. If only totals are sent to corporate headquarters, it may be more efficient to receive the information on paper and manually transfer the information into an electronic spreadsheet than to merge the information from a set of electronic spreadsheets.

Calculation Errors in Budget Spreadsheet

Another problem is calculation errors in the budget spreadsheet. This is very common if a customized spreadsheet has been created to budget the particular needs of the corporation. Typical errors are totals that do not include all the detail-level expenses listed in the budget, different pages of the budget that do not summarize properly into the summary page of the budget, seasonality factors for each month that do not create correct results by month, and dollar amounts that do not change when the overall activity level of the corporation (usually revenue) changes.

Manually Recalculate the Spreadsheet.   The most painful way to ensure the accuracy of the budget model is to recalculate the entire model with a calculator. This involves footing all column totals, tracing percent changes from the initial percentage to the resulting revenue or expense change, and tracing page totals forward to the budget summary page. This is a time-consuming project that can be rendered useless by subsequent changes to the model. The best way to preserve a fully tested model is to save a copy of the budget and store it in a safe location for recall in case subsequent versions of the budget prove to be excessively flawed.

Reduce the Number of Accounts.   If there are fewer account line items, then there are fewer accounts to summarize and fewer totals that will be wrong. The controller should review the activity in each account at the end of each year and determine which small-volume account numbers should be merged into some other account number. By pursuing this strategy every year, the number of accounts can be kept to a manageable level.

Summarize Small-Dollar Departments.   If a department has a negligible dollar total, then it should be allowed no more than a single line item in the budget. This prevents summarizing errors for that department—if there are no line items to summarize, then there can be no calculation error.

Shrink Monthly Budgets into Quarterly Budgets.   The typical company budget contains a budget for each month of the year and therefore twelve chances to create errors. The number of calculations in the budget model can be reduced to one-third of the previous level simply by switching to a quarterly budget system. If the company uses a budget versus actual reporting system for each month, it can divide each quarterly budget amount by three to derive its monthly budget for reporting purposes.

Change the Color of User-Defined Cells.   The typical budget contains an array of cells, some of which are calculations and some of which require manual entries. The difference between these two types of cells may not be clear to someone who is trying to rapidly enter modifications to the budget. To avoid the problem of having a calculation cell erased, the manual entry cells can be colored-coded in most major electronic spreadsheet programs, which gives the budget analyst a visible warning about the type of cell on the screen. To be doubly sure that calculation cells will not be deleted, they can also be protected from modification with passwords.

Purchase an Off-the-Shelf Budget Program.   Most budget analysts want to create their own budget models, but it is possible to purchase an off-the-shelf budget program that has been pretested and (in theory) contains no calculation bugs. These models have the added advantage of calculating cash flows and balance sheets, which a custom-designed budget model may not accurately generate (or even include).

Inconsistent Changes to Variables

A final and very common budget error occurs when multiple iterations of the budget are created. The error occurs when one variable is changed without corresponding changes to related variables. For example, if the revenue amount is changed, the budget should automatically change in a number of other areas, such as the amount of production supplies, headcount, and salesperson travel expenses. If fixed expense amounts are used, then the budget analyst must manually scroll through the budget model and manually change the variable expenses. If there are a large number of variable expenses, then it is likely that the budget analyst will not alter all the expenses, resulting in inaccurate budget results.

Use Percentage Calculations for Variable Expenses.   For a key variable expense such as the cost of goods sold, the expense should be directly linked to the revenue level with a calculation, rather than recording the cost of goods sold as a fixed cost. Percentage formulas are easy to include but should be identified with a colored cell to highlight their existence.

Cluster All Key Variables at the Front of the Model.   If all key variables are listed in one screen of the budget model, then the budget analyst can review them all at the same time. This is crucial when one variable has been changed, since this may require a change to another variable. Key variables that should be listed in one place are the total revenue amount, direct labor percentage, materials percentage, overhead rate, headcount in various key departments, and tax rate.

In summary, a few simplifications of the budget model will keep the number of budget-related errors to a minimum. Since many of the changes involve shrinking the size of the budget, there is a side benefit of being able to update the budget much more quickly than would be the case with more account numbers, detailed budgets for insignificant departments, manually altered variable costs, and budgeting for a full 12- or even 13-month budget year.

COST/BENEFIT ANALYSIS

This section contains an overview of how to write cost/benefit analyses for implementing a video conferencing system and streamlining the budget model to speed the budget creation process, as well as an overall cost/benefit analysis that incorporates all other changes to the budget cycle. In these examples, expected revenues and costs are as realistic as possible.

Video Conferencing

The SpaceTech Company constructs customized parts for satellites to be launched by many countries. The parts are constructed in eight locations around the world. To speed up the budget process and save travel costs, the budget director chooses to implement video conferencing. She finds that, in the previous year, the company sent two people from each location to the central headquarters to discuss the budget, and sent everyone to headquarters twice. The average travel cost per person, including air fare, hotels and meals, is $1,800 per trip. The cost of using video conferencing is $55,000 to set up a system at each location, plus $30 per hour of telephone time to transmit signals. A total of 32 hours of transmission time will be needed to complete the budget. The company depreciates this type of equipment over five years. An alternative approach is to travel to local rented video conferencing facilities, which can be rented for $150 per hour, plus $40 per hour for telephone charges. Is it worthwhile to implement video conferencing, and how?

Solution. The cost savings are in reduced travel and entertainment costs. There are two ways to implement video conferencing—either by rental or by purchase of the equipment.

Cost of Travel to Meetings (No Video Conferencing)

No. of locations needing travel	7
No. of people/location	× 2
Total no. of people traveling	14
Trips/person	× 2
Total trips	28
Cost/trip	× $1,800
Total travel cost	$50,400

Cost of Purchasing Equipment for
Video Conferencing

No. of locations needing equipment	8
Fixed asset cost/location	× $55,000
Total capital cost	$440,000
5-yr. depreciation rate	÷ 5
Total depreciation/yr	$ 88,000
No. of transmission hours	32 hrs
No. of locations transmitting	× 8
Total transmission time	256 hrs
Telephone cost/hour	× $30
Total transmission cost	$ 7,680
Total purchased equipment cost	$ 95,680

Cost of Renting Equipment for Video Conferencing

No. of transmission hours	32 hrs
No. of locations transmitting	× 8
Total transmission time	256 hrs
Equipment and telephone cost/hour	× $190
Total rented equipment cost	$48,640

Thus, with travel costs of $50,400 and equipment rental costs of $48,640, it is reasonable to implement video conferencing with rental video facilities. The purchased equipment option is too expensive to implement, unless other uses can be found for the equipment when it is not being used for budget purposes.[1]

Simplify the Budget Model

The controller at fast-growing InterMode, a maker of model trains, is concerned that the budget process is taking too much of the financial analyst's time each year. By altering the budget model to include fewer accounts, varying some expenses directly as revenue fluctuates, and cutting back reporting periods to quarterly, it should be possible to reduce the hours that the analyst devotes to the

---

[1] The trouble with using rented facilities is that there may not be any facilities close to some company locations. Also, the productivity cost of having key employees travel is a difficult number to quantify but can be included in a cost/benefit analysis. In addition, the central video conferencing facility must be able to accept video feeds from multiple locations in order to conduct a conference with people from multiple locations—this is expensive. Finally, a new set of video conferencing tools are becoming available that cost about $1,000 per computer and will allow conferences to be conducted from individual workstations. When these personal video systems become widespread, the economics of using video conferencing as a travel alternative will become much more favorable.

budget. Further examination reveals that, in the previous year, the budget went through four iterations that required reentry of information into the budget for all four iterations. The analyst estimates that it took two days to revise the budget on each occasion. By making the preceding changes, the analyst should be able to cut the reentry time in half. However, it will require two days of work to alter and test the budget model to ensure that the simplifications do not cause any errors. The financial analyst earns $45,000 per year. Should the InterMode company alter its budget model?

Solution. The controller must know if the cost of altering the budget will generate a payback by reducing the work load to the financial analyst.

Cost of Streamlining the Budget Model

No. of days to alter budget model	2
Daily cost of analyst	× $173
Total cost of new budget model	$346

Benefit of Streamlining the Budget Model

No. of budget iterations/yr	4
No. of days to alter budget model	× 2
Total days/yr to alter budget model	8
Time savings from new budget model	× 50%
No. of days saved by new budget	4
Daily cost of analyst	× $173
Total savings from new budget model	$692

With costs of $346 and savings of $692, the budget alteration cost achieves a break-even after only two budget iterations and saves money thereafter. The streamlining project should proceed.[2]

Reduce Budget Review Time

The managers of Gregorian, Inc., producers of business calendars, are bothered by the amount of time they spend preparing the budget each year. The CFO recommends a number of changes to the process, including setting available funding

---

[2] The costs and savings are not "hard," since the financial analyst is presumably on salary and will not be paid less if the budget is completed in less time. However, it does allow the analyst to work on other projects that may save the company money. Also, some budget models are extraordinarily complex and may require far more revision time than the interval noted in this case study. Usually, shrinking to quarterly budgeting periods and reducing the number of accounts is quite easy to implement in the model, while installing a flexible budget is much more time-consuming.

levels in advance, eliminating detailed expense reporting until the high-level budget is complete, prioritizing projects based on how they relate to corporate goals, and linking the budget to a performance measurement and reward system. These changes are almost free to implement, since the only cost is three days of the CFO's time to construct a new budget flowchart and a reward system. With these changes, the CFO feels, the management team can cut the number of budget iterations from seven to three, though the typical three meetings per iteration will probably increase in duration from three to four hours. There are six members of management on the budget committee, and their pay averages $75,000. The CFO is one of the six members of the budget committee. Should these changes be implemented?

Solution.   The number of hours saved by reducing budget iterations must be offset against the time of the CFO to create the budget flowchart and reward system.

Cost of Implementing Budget Review Changes

Days of CFO time needed	3
Cost/day of CFO	× $288
Total cost of implementing changes	$864

Benefit of Implementing Budget Review Changes

No. of budget iterations eliminated	4
Meetings/iteration	× 3
Total no. of meetings eliminated	12
No. of people attending each meeting	× 6
Total savings from people attending meetings	72
Time/meeting	× 3 hrs
Total no. of attendance hours saved	216 hrs
Less: increased length of remaining meetings	− 54 hrs
Net no. of meeting hours saved	162 hrs
Cost/hour/manager	× $288
Total savings from reducing budget review time	$46,656

With costs of $864 and savings of $46,656, there is an overwhelming argument in favor of implementing the budget streamlining project immediately.

In summary, the best way to justify changes to the budget process is to quantify the cost of management time needed to produce a budget under the current system and then to project the time savings that accrue from implementing the streamlining suggestions advocated in this chapter. Since most of the changes have a minimal cost, the primary analysis becomes focused on the hours and labor cost of management time saved by implementing the changes. Also, video conferencing can be implemented in some cases for companies with far-flung operations, though the savings may not be evident unless other uses can be found for the equipment or if rental facilities are located near the subsidiaries.

REPORTS

No periodic reports are needed besides the budget. However, the budget process can be assisted by a brief budget procedure that outlines due dates and responsibilities. This section includes an example of a very brief budget procedure for a company with no subsidiaries (adding subsidiaries can greatly lengthen the procedure). Also, this chapter has discussed a variety of changes in the traditional budget model, in terms of fewer line items and reporting periods, using a high-level model for determining the corporate direction. While such a revised budget model should be clear, the composition of a high-level decision-making model may not be as evident. Consequently, this section also includes an example of a summary management budget model.

A budget procedure should include, at a minimum, the nature of various budget deliverables, the dates when they are due, to whom they should be sent, and who is responsible for sending them (see Figure 7.3).

AJAX SYMPHONIC RECORDINGS, INC.
1998 Budget Procedure

Deliver Budget to Board First Monday of December
Send Deliverables to CFO   **Page 1 of 1**

Step	Date	Responsibility	Deliverable
1	10/05	Budget Committee	Review strategic direction
2	10/15	Sales Vice President	Present revenue plan
3	10/20	Marketing Vice President	Present marketing plan
4	10/23	Production Vice President	Present production plan
5	10/28	Eng. VP, MIS VP, CFO	Present engineering, administration, and MIS plans
6	11/02	Production Vice President	Present facilities plan
7	11/09	CFO	Present capital expenditures plan
8	11/11	CFO	Present financing plan
9	11/20	Budget Committee	Budget reiteration meeting
10	11/25	Budget Committee	Budget reiteration meeting
11	11/30	Budget Committee	Final management review meeting

FIGURE 7.3   Budget procedure.

The following are some observations regarding the budget procedure.

Length.   The procedure does not need to be very long to lay out a reasonable listing of due dates and responsibilities.

Job Titles.   The procedure does not list the names of individuals who are responsible for various deliverables, since they may switch to new jobs. Instead, to keep the accounting department from having to reissue the procedure every time a new person becomes involved in the process, the job title that is responsible for each deliverable is listed.

Procedure Date.   It is important to list the date of the procedure, so that users will be able to compare different issues of the procedure and know which is the most current.

Procedure Page.   It is important to list the number of pages in the procedure, in case an employee does not complete a budget step by being unaware that a page is missing.

Task Dates.   It is difficult to reissue a budget procedure year after year without making changes to it, because the due dates may not fall on work days every year. Thus, at least the dates must be changed or listed as specific work days of the month (e.g., the second Tuesday of November).

Summary Management Budget Model

The focus of a summary management budget model is the information summarized in its tables. For this example, Tables 7.3–7.12 display the essential information needed by management to determine changes to the budget. Since there are very few expense line items in this model, any results from the model will be approximate. Nonetheless, it can be used to determine the viability of a new set of management assumptions in a few moments, which allows management to alter the budget during its high-level budget meetings. Being able to model a number of changes quickly allows the management team to speed up the budget process and keeps the budget support staff from having to recalculate the budget details an excessive number of times. This model would have to expand in size if the company had many more departments, or subsidiaries; the example assumes just a few departments and no subsidiaries. Also, more detail may be necessary depending on the variety of products sold, especially if the gross margin on different products varies appreciably—this would require an increased number of line items in the revenue and cost of goods sold sections of the model.

**TABLE 7.3   Summary Management Budget Model:
Revenue and Cost of Goods Sold**

Revenue	
Annual	$13,000,000
Monthly	1,083,333
Cost of goods sold, per revenue dollar	
Materials	$ .217
Freight	$ .03
Cost of goods sold, per year	
Materials	$ 2,821,000
Freight	390,000
Total	$ 3,211,000

TABLE 7.4   Summary Management Budget Model: Department Payroll Cost

Department	No. of Employees	Average Salary	Total Salary Cost	Benefits Percentage	Total Payroll Cost
Production	120	$32,000	$3,840,000	20%	$4,608,000
Sales	12	68,000	816,000	22	995,520
Marketing	2	52,000	104,000	22	126,880
Engineering	18	55,000	990,000	22	1,207,800
MIS	3	43,000	129,000	22	157,380
Administration	7	29,000	203,000	22	247,660
Total	162		$6,082,000		$7,343,240

**TABLE 7.5   Summary Management Budget Model: Department Cost
That Varies by Number of Employees**

Department	No. of Employees	Cost per Employee	Total Employee-Variable Cost
Production	120	$ 1,100	$132,000
Sales	12	15,800	189,600
Marketing	2	1,500	3,000
Engineering	18	1,250	22,500
MIS	3	3,100	9,300
Administration	7	2,200	15,400
Total	162		$371,800

TABLE 7.6   Summary Management Budget Model: Department Cost That Varies by Revenue Amount

Department	Cost per Revenue Dollar	Total Revenue-Variable Cost
Production	$.03	$ 390,000
Sales	.025	325,000
Marketing	.005	65,000
Engineering	.008	104,000
MIS	.01	130,000
Administration	.007	91,000
Total		$1,105,000

TABLE 7.7   Summary Management Budget Model: Department Fixed Cost

Department	Total Fixed Cost
Production	$108,000
Sales	25,000
Marketing	8,000
Engineering	62,000
MIS	80,000
Administration	42,000
Total	$325,000

TABLE 7.8   Summary Management Budget Model: Profit

Revenue	$13,000,000
Cost of goods sold	3,211,000
Department payroll cost	7,343,240
Department employee-variable cost	371,800
Department revenue-variable cost	1,105,000
Department fixed cost	325,000
Total cost	$12,356,040
Net pre-tax profit	643,960
Taxes (38%)	244,705
Net after-tax profit	$ 399,255

**TABLE 7.9 Summary Management Budget Model:
Working Capital**

Accounts receivable	(45 days)	$1,250,000
Inventory	(10 turns)	627,400
Total		$1,877,400
Accounts Payable	(30 days)	522,833
Net working capital		$1,354,567

**TABLE 7.10 Summary Management Budget Model:
Capital Spending**

1	Manufacturing equipment upgrade	$ 25,000
2	Quality monitoring system	80,000
3	On-line receiving system	15,000
4	Bill of materials software	2,500
5	Material requirements planning (MRP)	5,000
6	Clean room construction	20,000
7	New lathe	12,000
8	New glass cooling tank	18,500
	Total	$178,000

**TABLE 7.11 Summary Management Budget
Model: Funding Available**

Working capital required	$1,354,567
Capital spending required	178,000
Total	$1,532,567
Cash flow from profit	399,255
Depreciation	210,000
Total	$ 609,255
Funds required	923,312
Funds available	1,000,000
Excess funds available	$ 76,688

**TABLE 7.12 Summary Management Budget Model:
Performance Measurements**

Department	Goal	Measurement
Sales	Sales/salesperson	$1,083,333
Marketing	Variable expense/employee	1,500
Production	Variable expense/employee	1,100
Engineering	Variable expense/employee	1,250
MIS	Variable expense/employee	3,100
Administration	Variable expense/employee	2,200

The following line items appear in this model:

Revenue (Table 7.3).   A single revenue amount per year is listed. This is an important number, because many of the expense amounts listed further on in the model will change automatically as the revenue level changes. Also, the revenue amount per month is listed. This is an automatic calculation and allows management to view the sales level that must be achieved every month to achieve the annual revenue.

Cost of Goods Sold (Table 7.3).   This line item only includes materials and freight costs; overhead costs are placed in the department cost categories, and the production labor cost is noted in the department labor category. Production labor tends to be relatively fixed, especially if the work force is highly skilled. Thus, rather than calculate production labor cost as a percentage of revenue, it should be listed as subject to change only if there is a specific hiring or layoff decision by management. The material and freight costs are calculated as percentages of revenue.

Department Costs (Tables 7.4–7.7).   Payroll costs are listed first, since they are usually the largest department costs, and then costs that vary with number of employees, for instance, telephone and travel costs in the sales department. Next are listed costs that vary directly with changes in revenue level, such as production supplies in the production department and commissions in the sales department. Finally, fixed costs, which do not vary at this revenue level or with number of employees, are listed. The portion of the fixed costs total ($325,000) attributable to depreciation and amortization is $210,000.

Profit (Table 7.8).   The profit is a calculation based on revenue less costs and taxes. The profit number also appears in the funding table as "cash flow from profit."

Working Capital (Table 7.9).   The days of accounts receivable and payable as well as expected inventory turns are noted as variables, and the amount of working capital required is calculated based on the revenue and cost of goods sold. In calculating the inventory turnover rate, the cost of goods sold includes the cost of materials, freight, production pay, and production fixed and variable costs. Working capital required, plus the capital spending amount, yields the total amount of funding required for the upcoming year.

Capital Spending (Table 7.10).   The funding required by all company projects for the upcoming year is listed in this table. The projects are prioritized in order of how well they support the company's strategic goals for the upcoming year. The capital spending total, together with working capital requirements and profits, yields the total funds required for the upcoming year.

Funding Available (Table 7.11). The maximum amount of funds available (level of debt approved by the company's owners) is listed next to the amount of funds required. This allows management to compare the amount required by their model to the amount actually available. A shortfall would call for recasting the model.

Performance Measurements (Table 7.12). A key set of performance measurements can vary with the size of the budget to inform management of the goals that must be achieved to ensure that the budget model will succeed. These performance measurement calculations will vary automatically as the levels of revenue and expense change.

As shown, this management budget model summarizes the important budget information in tabular form. Written comments should be kept to a minimum. If required, they can be included as notes attached at the bottom of the pertinent tables or as short introductory or summary text.

In summary, a brief budget procedure is useful for informing employees involved in the budget effort of the due dates and deliverables required of them. This procedure keeps key information from being delivered late. Also, the summary management budget model, though abridged, is useful for determining the balance of cost and revenue needed by the company to achieve its profit goals within specific funding constraints. The summary model allows a quick review of revenue and expense activity for the upcoming year, which reduces the time managers must spend in the budget process. It also shrinks the time required by the budget support staff to complete the budget at a detailed level.

TECHNOLOGY ISSUES

There are few technology issues that influence the budget process, but video conferencing and electronic mail can be used to rapidly send budget information to budget team members, resulting in the removal of mail float and "people float" (employees' traveling to a central location for a meeting), thereby speeding the entire budget process.

Electronic spreadsheets can be linked to electronic mail messages and sent to employees throughout the company using such e-mail packages as Microsoft

Mail and CC:Mail. Also, files can be linked to electronic mail messages that are sent within the networks of the large national services like America Online or CompuServe. This is not possible if an employee has constructed a budget on an electronic spreadsheet and then tries to link it to a mainframe-based electronic mail system. Most of the mainframe systems do not allow file linking. The benefit of sending files by electronic mail is that employees can use the transmitted files for "what if" analysis as well as to send modified budgets back to the person who is consolidating budget information from numerous contributors.

Video conferencing is used to transmit images of employees between locations. This keeps them from having to travel to a central meeting location to discuss the budget and saves considerable employee travel time and cost. Video conferencing systems send motion pictures of employees over telephone lines to other locations, which are also transmitting video and audio signals back to the first location. Several cameras can be used to transmit multiple signals, usually of the participants, and more recently of documents that are either positioned beneath an overhead camera or clipped to a corkboard or whiteboard. The video conferencing room sometimes contains a computer and a fax machine, so that documents under discussion can be modified and sent by fax to the other locations. The following companies can provide more information about video conferencing equipment.

Alpha Systems Lab MegaConference	(800) 576-4275
AT&T Vistium	(800) 843-3646
Creative Labs ShareVision	(800) 998-1000
FarSite	(606) 245-3500
Insitu Conference	(617) 720-0821
Intel ProShare Video System	(800) 538-3373
LiveShare Plus	(508) 762-5000
PictureTel Live	(800) 716-6000
TALKShow	(415) 254-9000

A very recent innovation that can be used in conjunction with a video conferencing system or in a stand-alone mode is the joint modification of documents over computers that are located in separate locations. Under this system, a file is stored in one computer but can be accessed by another computer by modem, which dials directly into the first computer to access the file. The two users can then talk on the telephone about specific parts of the documents while the files are on-screen in front of both of them, though this requires a second phone line. Changes made to the file are immediately apparent to both users. The following companies can provide more information about remote file access.

Carbon Copy	(800) 822-8224
Close-Up	(805) 964-6767
LapLink Remote Access	(800) 343-8080
Reach Out Remote Control	(800) 677-6232

In short, technology can be used to bring budget information to users very rapidly, rather than having users come together to a single location to review the information, which is much more expensive and time-consuming.

MEASURING THE SYSTEM

The usual way to measure a budget is how closely budgeted expenses and revenue match actual expenses and revenue. However, very few companies have devised measurement systems that track the efficiency of the budget process. There are several ways to track budget efficiency.

The first measurement method is to track the time period covered by the budget process. This is a very simple measure involving recording the start date of the process and the completion date of the last budget task, which is usually the entry of the completed budget into the accounting software. This information is useful for comparing the duration of the budget cycle to that of previous years.

The next measurement method is tracking the total hours expended on the budget cycle. This is more complex, because it involves asking (in most cases) salaried workers, including senior managers, to track their time with time cards. However, estimates can be used as well. This measurement is somewhat better than tracking the overall duration of the budget cycle, since it gives some indication of the actual working time spent by the management team on the budget. A further refinement of this measurement is to multiply the hours worked by the burdened pay level of the people working the hours. Since some of the people involved in the process are the highest-paid members of management, this can provide a very illuminating picture of the price of creating a budget, and particularly the price of creating extra iterations of the budget.

Another measurement method is the number of accounts listed in the budget. Since the budget is easier to construct when there are fewer accounts, this measurement essentially tracks the simplicity of the budget. The measurement is simple enough to derive—just print out the chart of accounts at the beginning of the budget project and count the total number of accounts listed there. This information can be reported with the monthly financial statements in order to highlight the continuing need to keep the number of accounts down to a practical minimum.

Several performance measures can be used to track the efficiency of the budget process. These vary in simplicity of calculation, with tracking the hours of management time being the most difficult. The most useful measurement is the overall cost of the budget process, which requires tracking the time of individual managers who work on the budget process as well as their cost per hour. This information provides a detailed look at the need to shrink the number of budget iterations, which is where the cost of the budget cycle can expand the most.

IMPLEMENTATION CONSIDERATIONS

The budget changes advocated in this chapter are generally greeted warmly by management, since less time is required to create and modify the budget, allowing management more time to perform other tasks.

However, some changes relate to shrinking the budget model, such as reducing the number of account line items, using flexible budget formulas, or cutting back to quarterly budgeting from monthly budgeting. When a budget model is revised, the budget analyst must assume that errors have been introduced into the model; to correct the situation, the model should be manually reviewed to ensure that all line items are added into the totals, that quarterly amounts add up to annual totals, and that subsidiary schedules can be traced forward to the summary budget.

Another implementation issue is determining the amount of funding available in advance. This is the responsibility of the CFO, who may cause problems by trying to present too many options involving changes of ownership or considerable increases in the level of debt. The best way to avoid this is to have the owners (either specific owners if the company is privately held, or the board of directors if publicly held) establish funding guidelines in advance. These guidelines should specify the level of debt that the owners are comfortable with (usually expressed as a debt/equity ratio) and the level of equity dilution that will be tolerated by issuing more shares. These guidelines constrain the CFO within fairly narrow boundaries and allow a maximum funding amount to be presented quite rapidly.

Also, if management decides to modify the budget model by determining the company's cost drivers, then sufficient time must be allocated for a cost analyst to review the company's costing structure. If the company has a production process rather than being simply a service or retail organization, then this costing review may take several months. A large firm may require a team of cost accountants to determine this information. No matter how large the firm, the cost of determining the nature of cost drivers must be budgeted before the work commences.

If a budget manual is created, it may be ignored by the budget personnel of subsidiaries and other departments unless there is an introduction to the manual, such as a training videotape or a personal training session conducted by the chief budget analyst. This process gains greater acceptance of budget procedures and ensures a smoother budget process than would otherwise be the case.

The issue that will generate the most interest from management is the determination of performance measurements and related reward systems, since this results in payments to the management team. Any modification of these measures by a self-interested management team can be avoided by restricting input to this stage of budget construction to the top management group only. It may be sufficient to let the chief human resources executive and the CEO determine the reward system. However, when presenting the budget to those two people, the controller should also give them a suggested list of performance measures; since the controller must track the performance measures, it is appropriate to suggest which measures to implement.

In summary, a number of implementation issues must be considered to ensure that the new budget is properly constructed. The budget model must be presented with adequate procedural training and debugged; guidelines given to the CFO for deriving a maximum funding level; and performance and reward system development restricted to the top management team. If all these problems are considered during the budget process, the chance of creating a successful budget is improved.

SUMMARY

The budget process is unlike many of the other systems discussed in this book. It involves no transactions and does not require data entry or the transfer of assets. Because of these differences, the budget process has little need for controls (only controls later in the process to ensure that the budget is being followed). Also, it cannot be appreciably improved with technology-based efficiency enhancements. However, the process can be streamlined by concentrating the budget effort into the front end of the process in order to avoid multiple iterations of the budget for detail-level expenses that should not even be derived until the high-level budget has been determined. The process can also be improved by streamlining the budget model itself—by cutting back on the number of accounts, shrinking the number of periods, and listing small-expense departments with a single summary expense line item. The overall process can be further improved by distributing a budget manual that clearly describes the process and defines who is responsible for various stages of the budget cycle and when deliverables are due.

By implementing all or some of these changes, a quality budget can be produced more quickly than had previously been the case, thereby giving the management team more time to devote to other tasks.

REFERENCES

Fisher, Anne B. "Is Long-Range Planning Worth It?" *Fortune,* April 23, 1990, pp. 281–284.

Henkoff, Ronald. "How to Plan for 1995." *Fortune,* December 31, 1990, pp. 70–79.

Knight, Ray A., and Lee G. Knight. "Planning: The Key to Small Business Survival." *Management Accounting,* February 1993, pp. 33–34.

Lubove, Seth. "It Ain't Broke, But Fix It Anyway." *Forbes,* August 1, 1994, pp. 56–60.

Mintzberg, Henry. "The Rise and Fall of Strategic Planning." *Harvard Business Review,* January/February 1994, pp. 107–114.

Schmidt, Jeffrey A. "Is It Time to Replace Traditional Budgeting?" *Journal of Accountancy,* October 1992, pp. 103–107.

Umapathy, Srinivasan. "How Successful Firms Budget." *Management Accounting,* February 1987, pp. 25–27.

Zarowin, Stanley. "Staying in Touch." *Journal of Accountancy,* December 1995, pp. 55–59.

PART TWO

Other Topics

Part Two of this book examines a number of areas that are needed to support the improvement of the functions discussed in Part One. For example, Chapter 8, Electronic Data Interchange, explains how to install an EDI system and notes a number of issues that will arise during implementation. The information in that chapter can be used to install the EDI systems recommended in several chapters in Part One.

Chapter 9, The Quick Close, brings together recommendations made in previous chapters to show how the accounting ledgers may be closed more quickly. Chapter 10, Process Documentation, demonstrates how to chart an existing process and document process move and wait times. Finally, Chapter 11, Effects of Change on Employees, delineates the "people problems" that arise when change is implemented.

In summary, the goal of the material in Part Two is to provide the reader with backup information that will assist in bringing about the innovations recommended in Part One.

8

ELECTRONIC DATA INTERCHANGE

The traditional means of communication has been with paper documents. Then some transactions were relayed over the phone, usually with a paper record of the verbal transaction. Then phone lines were used to transmit fax information. Incremental improvements occurred when lockboxes reduced the mail float. Overnight delivery services have also helped to speed the transfer of documents. Despite all these improvements in the transmission of information, the computer systems of one company must still convert information into a new format (usually paper), which must then be entered into the computer system of the recipient. This rekeying of information means that transferred information is subject to keypunching errors. The keypunching errors have been reduced by creating data entry software with built-in edit checks as well as duplicate data entry systems that cross-check two sets of entered data for errors (though requiring twice as much labor to input the information). Also, computer systems were created that passed information *within* a company from one computer application to another electronically. Recently, optical character recognition systems have been invented that scan incoming documents and translate them into electronically stored documents. Despite all these improvements, there is still a problem with seamlessly transferring information between companies. Electronic data interchange (EDI) solves this problem for some transactions.

This chapter provides an overview of EDI, its history, a practical guide to implementing an EDI system, information about the more common EDI transactions, and references for gathering more information about EDI.

DEFINITION

Electronic data interchange is the transmission of information between computers in a strictly defined, prearranged format. The information flow requires no paper. Previously, a paper transaction between companies involved entering information into a computer, printing the result, sending it to the receiving company by fax or mail, entering the information into the receiving company's computer, and finally processing the information. EDI simply sends the information from one company's computer to the other without the rekeying or mailing steps. Rekeying is costly in labor and introduces delays and errors.

The prearranged format used for transactions may vary by industry or country. For example, the grocery and pharmaceutical industries have developed transaction formats that are precisely defined for the needs of those industries. However, most companies adhere to the ANSI X12 standards, which are designed to be applicable across industry lines. Companies dealing in the international arena use the EDIFACT standard, which is similar to the ANSI X12 standard.

HISTORY

EDI began in the 1970s when several industries independently formulated data format standards known as transaction sets. The definition of industry-specific transaction sets meant that different industry sectors could not communicate. Therefore, in 1979, the American National Standards Institute (ANSI) formed the Accredited Standards Committee on EDI-X12; its charter was to formulate generic EDI standards that would be useful across industry lines. The committee's members come from many companies across a number of industries as well as from government. The committee has now formulated over 100 ANSI standards that define the data formats for nearly every transaction in the business cycle, from the initial request for quote to the final payment.

EDI has caught on primarily through the efforts of large companies who require their trading partners to use EDI for some or all trading transactions with them. Many large companies have proprietary systems; they are linked directly to their trading partners with custom software as well as special hardware and communications equipment. The problem with a proprietary system is that trading partners are locked in to a customer's or supplier's system and must set up an entirely different set of hardware and software to correspond with a different trading partner. In government, the Department of Defense has been a strong supporter of EDI and now uses it for purchasing, invoicing, payments, requests for bids, and responses to requests for bids.

Proprietary systems evolved into industrywide standards that were fostered by industry trade groups. These led to specific standards for such industries as trucking and pharmaceuticals that were tailored to the needs of each industry but required modification to be used across industry lines. The following are some industry-specific EDI standards.

EDI Acronym	Industry Using the Format
WINS	Warehousing
TDCC	Transportation
UCS	Grocery
ORDERNET/NWDA	Pharmaceuticals

The next step up from industrywide standards was a set of standards that would be relevant across industry lines. The standards that were developed tend to be all-inclusive, so that the data needs of all industries are met. However, the excessive complexity of the standards means that any single industry does not have a need for all the data elements required by the standards. Consequently, many industries have now adopted the cross-industry EDI transaction standards but have elected to use only a subset of the data elements required by each full standard transaction. Similarly, each industry uses only subsets of the total number of EDI transaction sets. For example, the stow plan transaction is used by the transportation industry to determine the exact location of a shipment on an ocean carrier. Other industries do not need this transaction.

The final stage in the development of EDI was its expansion to the international arena. For cross-country transmissions, the set of standards used is called EDIFACT (Electronic Data Interchange for Administration, Commerce, and Transport). The first EDIFACT transaction set was approved in 1987, so this is a fairly new standard. Because the EDIFACT standards vary slightly from the standards developed by ANSI X12, users should be aware that integration of EDI into their existing software will require more work to ensure that both standards work seamlessly. Alternatively, if all of a company's business is located within the United States, then only the ANSI X12 standards can be integrated into the existing software.

What is the future of EDI? It has not yet gained acceptance throughout industry but will continue to take over from paper-based transactions and spread into the international trading area. It is most heavily used between trading partners who conduct a large volume of transactions. EDI will grow more slowly into the ranks of companies that have small volumes of transactions and that will therefore have less need for repetitive electronic transactions.

BENEFITS

The benefits to be derived from installing EDI fall into the following categories.

Improvement in Data Accuracy

The typical rate of error on entries into a computer is between 3% and 5%. Eliminating the rekeying of data avoids a large number of errors, which reduces the staff time needed to correct the errors.

Improvement in Cycle Time

The cycle time to exchange information about requests for quotes, quotes, purchase orders, invoices, and payments is considerable. EDI appreciably reduces this cycle time by eliminating the time required for paper documents to move between trading partners. The frequent transfer of order information between a company and its suppliers for JIT production is facilitated by EDI, since many orders per day can be sent to suppliers with nearly instantaneous transmission speeds and minimal cost per order. Planning data can also be sent by EDI, so that suppliers can plan their production schedules around the company's requirements.

Elimination of Mailing Costs

All costs associated with mailing documents to trading partners are eliminated. These costs include expenditures for envelopes and postage, and the labor required to stuff envelopes, affix stamps, and sort envelopes for bulk mailing.

Elimination of Mail Float

The time required to wait for delivery of documents by the postal service can be eliminated by EDI. However, most companies batch their transactions and send them in a cluster to the value-added network (VAN) rather than sending them to the VAN one at a time. If batches are only transmitted infrequently, then the mail float time has simply been replaced by an EDI transaction accumulation time. An example of mailing a transaction versus using EDI is shown in Figure 8.1.

Reduction of Safety Stocks

Safety stock is extra inventory kept on hand to keep a company from running out of stock during the period when an order has been placed for more inventory but has not yet arrived. Since EDI reduces the time required to place an order, the period during which safety stock is needed is reduced. Consequently, the amount of required safety stock goes down, which reduces working capital requirements.

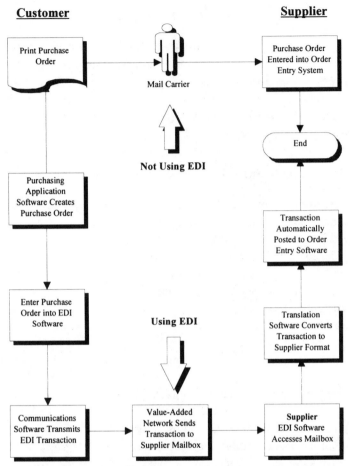

FIGURE 8.1 Mailing a transaction versus using EDI.

Elimination of Paper Handling and Storage

A manual system that sends paper documents to trading partners involves paper handling at many steps: when the document is created, sent to the trading partner as well as to any internal destinations, filed away, and retrieved for later purposes. Later, the document must also be sent to off-site storage, be tracked for eventual destruction, and finally destroyed. Clerical costs are associated with all these steps. An EDI system eliminates all of them. In addition, the space taken up by filing cabinets could otherwise be used as valuable office space by company employees. Another major problem with filing paper documents is that misfiled paperwork is extremely difficult to find; alternatively, computerized records are quite difficult to lose.

Electronic Forms

EDI can be enhanced by linking it to electronic forms at the sending company. For example, an electronic form for ordering office supplies can be set up for easy access by designated employees. They order office supplies on a form that resembles a preprinted order form. The information on the electronic form is then translated into the EDI purchase order format and automatically sent to a trading partner to have the order filled.

Automation of Additional Processes

The greatest savings involving EDI result when it is used as part of a greater scheme of automation within the company. For example, a purchase order can be automatically loaded into the order entry system, which automatically loads the order into the company's production schedule, which automatically sends an EDI transaction back to the customer with an expected shipment date. A fully automated system will also issue purchase orders if supplies are too low to build the product, issue picking tickets for collecting raw materials, and even recommend additional staffing for the time period in which the product is scheduled to be built. As another example, invoices can be received via EDI, automatically matched to purchase orders in the computer, creating an automated request for payment, which is then handled by electronic funds transfer (EFT). In both examples, EDI is only a small part of the total automation envisioned; thus, EDI can be used as the starting point of a chain of interlocking automation projects that results in minimal manual intervention (usually to correct exception items).

Based on this view of cost savings, there are some EDI transactions with low cost-effectiveness, such as a request for quote (RFQ)—it usually cannot be responded to automatically, especially if custom work is required, because there are attached notes that must be examined manually, thereby eliminating any savings due to automation. Again, with this view of cost savings, perhaps the most cost-effective EDI transaction is the purchase order, since it can be automatically linked to so many other processes.

If a company does not link its EDI system to its other computer systems, it must manually rekey all the data into and out of the EDI software. This type of system is essentially a fax machine. Since fax machines are cheaper than EDI systems, a company would be better off purchasing a fax machine.

The benefits to be gained from an EDI system can be negated by the sending of inaccurate information. No matter how good the EDI system may be, if the received information is incorrect or incomplete, it will be of no use to the receiving company.

In summary, the benefits accruing from a fully automated EDI implementation include the elimination of transaction-processing time because of the elimination of mail float as well as the reduction of costs due to reduced or eliminated filing, safety stocks, paper handling, and data rekeying.

WHEN TO USE EDI

Electronic data interchange should not be used for all company transactions, since the expense associated with converting some less frequent transactions to EDI may be considerable. EDI is most useful when the following conditions exist.

Clerical Costs Are High

Many businesses input large volumes of orders into their order entry systems every day. Clerks enter the orders into the computer. When benefits and payroll taxes are included, clerical costs are quite high. EDI transactions can be linked directly to a company's order entry system and bypass the clerical entry stage, thereby saving labor costs. Of course, any automated error-checking system will still require manual intervention to fix any flagged problems with incoming transactions.

Time Is Critical

If orders must arrive on time and be shipped quickly (such as in the grocery business), then EDI can be used to eliminate the mail float time.

Transaction Volume Is High

The cost of designing, purchasing, and implementing an EDI system is high and can most easily be justified by eliminating the manual labor associated with processing the highest-volume transactions. The highest-volume transactions can usually be found in the accounts payable and accounts receivable areas of a company.

Data Must Be Accurate

When the correct information absolutely must arrive at the trading partner in the correct format, so that it can be inputted directly into the receiving company's computer system, EDI would be useful.

There Are No Special Instructions

An automated review of an incoming EDI transaction will not work very well if the transaction is accompanied by a long list of notes. EDI works best when the information being sent will fit into a strict format and use a limited number of variables, so that error checking is simplified. In short, transactions full of narrative are poor candidates for EDI.

Transactions Cross International Boundaries

A considerable problem with international trade is the language barrier. A purchase order written in English may be mistranslated into another language, such as French. However, an EDI transaction uses codes to represent longer strings of information, and it is difficult to mistranslate a code. For example, a paper purchase order may contain a request for an air filter that is 2 inches thick. A recipient may have trouble translating *thick* since it can be defined as breadth, depth, or width. As a result, the ordering company may receive an air filter with the wrong dimensions. However, when an EDI system is set up, codes are agreed on and defined in advance. To return to the example, an EDI transaction for the air filter might simply list the product as code 9876-023. The recipient will then look up the code in a table that was predefined in cooperation with the sending company, which precisely lists the dimensions of the air filter. Thus, ordering with EDI codes tends to reduce language translation errors.

Trading Partners Require It

Especially in the retailing industry, customers who purchase in extremely large quantities are requiring their trading partners to establish EDI links or to risk losing all business with them. In these cases, a company has the choice of either losing a substantial piece of its business or of implementing EDI.

EDI is less useful when trading partner contacts are infrequent, orders are for custom products, trading partners change frequently, or order volumes are extremely low. The same reason for not implementing EDI applies to all these situations—the company will not be viewed as a key customer by its trading partners, and so they will not be willing to invest the time and money needed to establish an EDI link with the company. In addition, computerized error checking of incoming EDI transactions is difficult if not impossible when orders are customized; automated flagging of mistakes is much easier when incoming orders are simple and standardized, reducing the number of possible mistakes to a small enough number for a computer to handle.

In short, all prospective EDI conversion projects should be reviewed against the preceding points to see if the projects are viable.

AN EDI SYSTEM

An EDI system involves the interaction of internal software, hardware, the systems of trading partners, and the communications systems that link the trading partners. The following briefly describes each element of an EDI system.

Hardware

A personal computer can be used to send or receive EDI transactions. However, the main benefit gained from using EDI is directly linking electronic transactions to a company's application programs in order to avoid rekeying of data. Since PCs tend to be stand-alone, any received EDI transactions are usually printed out and rekeyed into other software. However, a PC can serve as a front-end to a mid-range or mainframe computer and transmit the information electronically to the more powerful computer for inclusion in the corporate database as well as receive the information from the larger computer for transmission purposes. Using a PC as a front-end to a larger computer is complicated; storing the EDI-related software on the same computer that houses the company's primary application programs is simpler and more efficient.

Translation Software

On the sending side of an EDI transaction, the translation software collects the data needed from various company databases to create the EDI record to be transmitted. This information may also be entered manually. It then converts the collected data into the format acceptable for transmission to the value-added network as an EDI transaction set (this portion of the software is usually purchased, not custom-designed). Conversely, translation software at the receiving end of the EDI transaction will split up the incoming record into data that are sent to the various corporate databases needed to process the transaction. Also, the receiving translation software should review the incoming EDI transaction for completeness and notify an operator of any incomplete data. The receiving translation software is a great cost-saver for the receiving trading partner, since it eliminates the manual conversion of date into the corporate databases.

This software is custom-designed and can take a substantial amount of time to develop. For example, an incoming purchase order may reference the part number used by the sending company; the translation software must look up the customer part number in a table that matches it to the receiving company's part number and replace the sender's part number with the receiver's part number. There are many EDI transaction sets, and many of them require data changes by the translation software, such as the one just described. Thus, every time the receiving company elects to begin receiving a new EDI transaction set, it must add to its translation software to accommodate any additional changes to the incoming data.

An issue for the system developer is the type of EDI standard to design into the EDI system. The ANSI X12 transaction sets are used domestically, whereas the EDIFACT standard (which includes extra data elements that are unique to international trade) is used for international transactions. Each transaction set has slightly different requirements regarding field lengths and numbers of data elements, so the system must be designed differently depending on which standard is to be used.

When the translation software is designed, the system developer must factor in the effect on those manual systems being replaced by the automatic uploading of arriving EDI transaction sets. For example, in the case of an invoice, a company's manual system will probably require matching with a receiving report as well as a purchase order, followed by storage in a carefully ordered filing system. When EDI transactions begin to arrive that bypass this system, the personnel responsible for the manual system will have concerns regarding access to the EDI information as well as controls over the information. Consequently, the old manual system should be thoroughly documented prior to installing an EDI system, so that the necessary elements of the old system can be incorporated into the new system.

Since trading partners are usually brought onto EDI over a long period of time, the manual systems must still be operational to deal with companies that are not scheduled to convert for some time. Also, some transactions may require manual processing for selected portions of each transaction, so some interface with manual systems may need to be designed.

There may be inquiries about EDI information being received, so an inquiry module is helpful for those employees who want to review incoming transaction sets. This module can include information about the content of each EDI transaction received within a user-specified time frame, sorted by transaction type.

When purchasing new applications software, it is worthwhile to require EDI translation software as part of the package; having it on the sending or receiving end of the applications software can save considerable design and programming time constructing such software.

Communications Software and Hardware

A company must have a modem and communications software to transmit EDI transaction sets to its trading partners. If it deals directly with its trading partners without an intermediary, then its communications protocol (software that checks for data receipt and acknowledgment) and line speed (rate of speed at which data are transmitted) must match that of the receiving trading partner. If there are many trading partners, this may entail setting up a large number of communications equipment variations. Using a value-added network can avoid this problem.

Value-Added Network (VAN)

A VAN collects incoming EDI transactions from one trading partner and stores them in an electronic mailbox for pickup by another trading partner. In cases when trading partners use different VANs, the receiving VAN will pass the EDI transaction to the VAN used by the other trading partner, so that the recipient trading partner can access EDI transactions from its own VAN. Multiple VAN usage is shown in Figure 8.2. Also, a VAN supports a variety of line speeds and communication protocols, so a company accessing a VAN would only need one communications configuration in order to conduct EDI transactions with all its trading partners, no matter what line speeds or communication protocols they use. An illustration of the problem with multiple line speeds used by trading

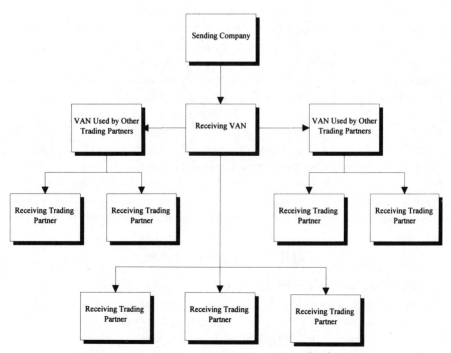

FIGURE 8.2   EDI transactions through multiple VANs.

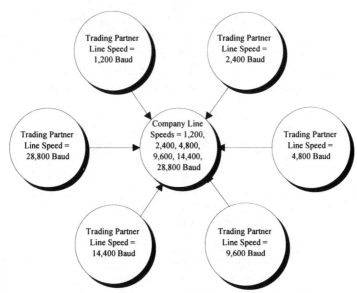

FIGURE 8.3   Need for multiple line speeds with different trading partners.

partners is shown in Figure 8.3. Also, a VAN operates 24 hours a day, so that transactions can be sent at any time. If trading partners were to send EDI transactions directly to each other, this might not be the case. In fact, transactions sent directly to a trading partner might be met by a busy signal, since another trading partner might be using the same phone line. An example of a typical transaction using a VAN is shown in Figure 8.4. Finally, corresponding directly with trading partners requires a single transaction with each one, which could add up to a large number of individual transactions in a day; however, by using a VAN, a company could access its electronic mailbox and send/receive all its EDI transactions with all its trading partners in just one telephone access.

A VAN can also send EDI transactions to the receiving party rather than wait passively for the receiving party to call in from time to time to pick up messages. The VAN can send its collected store of messages at a specific time of day or when a specific number of transactions have accumulated, or when a specific type of transaction set (e.g., a purchase order) arrives in the mailbox.

EDI Conversion Service

There may be some smaller trading partners who are unwilling to switch to EDI. In those cases, it is possible to have an outside service convert EDI transactions to those companies into faxes, and to convert incoming paper documents from them into EDI transaction sets. This service is expensive; before using it, one

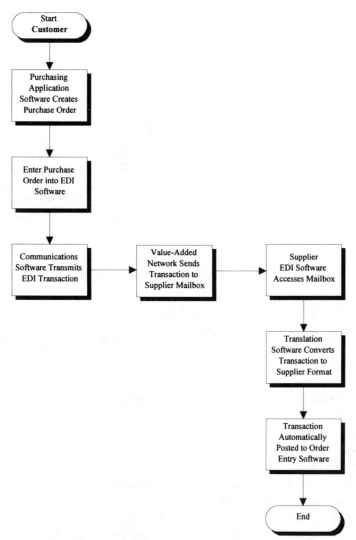

FIGURE 8.4   Typical EDI transaction using a VAN.

should compare the cost to that of maintaining a system in-house that continues to process paper transactions.

The EDI Record

An EDI record has three parts. First, there is the transaction header, which lists nonrepeating information for each record, such as an invoice or purchase order

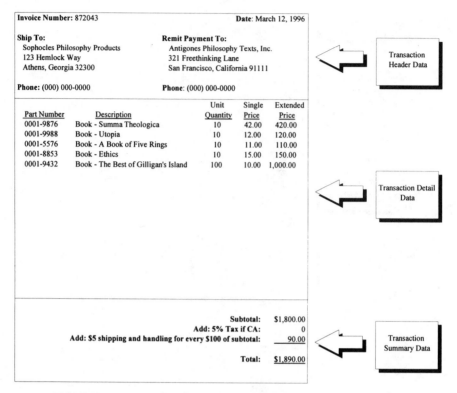

FIGURE 8.5   Paper invoice and corresponding electronic record.

number, the transaction date, and the customer identification number. Next is the transaction detail, which lists the subject of the transaction, such as items to be ordered, billed, or inquired about. Finally, the transaction summary lists any report totals, such as the total quantity ordered or the total amount billed. A paper invoice and the corresponding parts of an electronic record of an invoice are shown in Figure 8.5.

TYPICAL EDI TRANSACTION

One of the most frequently used EDI transactions is the purchase order transaction. The purchase order is used for the following example of the steps involved in processing an EDI transaction to help the reader understand the specific procedure to be followed.

1. Enter a purchase order into the company's computer system.

2. The EDI translation software reformats the purchase order into an EDI format.

3. Communications software sends the transaction by modem to a VAN.

4. The recipient's communications software retrieves the information from the VAN.

5. The recipient's translation software reformats the purchase order information from the EDI format into the format used by the recipient's order entry software.

6. After reviewing the order, a sales representative confirms that the purchase order can be handled (in a highly automated environment, the computer can handle this step; however, many companies using master production schedules would want to have a manual review of additions to the master schedule).

7. The order entry system creates a purchase order acknowledgment that sends a confirming EDI transaction back to the initiating company.

8. If no confirmation transmission is received by the company that sent the original EDI transaction set, then after a specific time period, follow-up action is taken to ensure that the original transaction set was received.

QUICK EDI INSTALLATION

Some companies will only want to install a basic EDI system as quickly as possible in order to meet the requirements of a large trading partner and will not be concerned about extending the benefits of EDI into other corporate applications programs. For those companies, here is a brief description of how to be up and running on EDI as quickly as possible.

1. Get the software. Obtain an EDI software kit by calling any of the major long-distance telephone companies, or review a copy of the EDI Yellow Pages, which is a listing of EDI service and hardware providers. To obtain a copy of the Yellow pages, call 800-336-4887.

2. Configure the system. Load the software into a PC that is connected to a phone line by a modem. Dial up the VAN to verify that you are properly set up on the system.

3. Notify trading partners. Contact your customers to inform them of your VAN identification number, so that they can use it to send information to you through the VAN.

4. Send a transaction. Enter a transaction set into your PC's EDI software, and send it to the VAN, having listed on it the identification number of the customer for whom it is intended.

5. Ensure that an acknowledgment is received. Once the customer receives the transaction set, an acknowledgment will be sent back to you through the clearinghouse. A good EDI software package will track the acknowledgment automatically for you, and tell you if the acknowledgment has not been received; if not, call your customer to verify receipt of the transaction set.

You are operational on EDI! You have met the requirements of your customer, but you have not received many of the benefits of EDI; instead, you are operating an expensive fax machine. Integrating the EDI transactions into other company computer databases and programs is where savings really occur.

IMPLEMENTATION CONSIDERATIONS

It is not the purpose of this text to provide a complete implementation guide for installing an EDI system. However, the following are some questions to consider during the design of an EDI system.

Does Upper Management Support the Project? A problem with EDI implementations is that the installation is seen as a technical improvement only. In actuality, it enhances the competitive position of the company by increasing the speed of transactions as well as reducing the number of errors in the system. Educating upper management regarding this issue is a challenge for the implementation team or project sponsor.

Has Responsibility for the Project Been Assigned to Anyone? The responsibility should be shared with managers from the areas being affected by the EDI implementation so they will not be able to blame someone else if the project is not successful. Also, their input and support ensures a higher degree of success.

Has a Cost/Benefit Analysis Been Completed? When conducting a cost/benefit study, be sure to review the current processes to be replaced, including all manual tasks, filing of documents (including filing of *all* copies of a document), and the total number of transactions to be replaced by EDI. In most cases, EDI will not eliminate all transactions, since not all trading partners will be using EDI. Therefore, the cost of maintaining systems to deal with paper transactions must be factored into the calculation. A key action is to contact trading partners who will have to switch to EDI in order to make the cost/benefit study show a profit, and verify that they are willing to participate in the project. A result of the cost/benefit study may be the minimum number of transactions needed to make the project pay for itself as well as the specific list of trading partners who must switch to EDI in order to achieve the break-even cost level. Finally, the study should include a best/worst estimate of the funding needed to complete the project.

What Is the Employee Attitude Toward Change? If previous technology-related projects have failed, then the project team should consider a "quick hit" approach whereby EDI is set up with a single trading partner as fast as possible to show that the system can work. Additional EDI partners and system features can be added later, once the system is operational.

What Changes in the Organization Will Be Caused by an EDI Implementation? The implementation team should work closely with a human resources department representative to determine the increases needed in technical personnel, reductions in clerical personnel, and retraining of employees assigned to handle exception transactions.

What Will Happen to Displaced Staff? After an EDI implementation has been completed, some staff will probably be displaced by the project. For example, the payables clerks who previously matched purchase orders to invoices will no longer have a job when EDI and additional process reengineering are used to match the documents electronically. What can be done with the staff? One option is to permanently shrink the department by letting the people go. However, a more attractive alternative is to retrain the employees in other tasks. There are two reasons for this. First, they have worked inside the company and therefore know enough about the staff, products, and procedures to be able to retrain into a new position much faster than a new person would be who was unfamiliar with the company. Second, the displaced staff has just seen the effect on efficiency of the EDI and associated systems restructuring, and may be willing to lead or participate in similar efforts elsewhere in the company.

What Education of Users Is Contemplated? Education should include not only the employees operating the software and hardware but also the people who will be handling exception transactions that are shunted to them by the system.

Have the Legal Aspects of EDI Been Reviewed? The key legal issue involving EDI is need for signatures on documents. A startup agreement should be reviewed by the company's legal staff that should be signed by every EDI trading partner. In addition, the company should switch to signed blanket purchase orders, with purchase order releases using EDI that do not require signatures.

Have Trading Partners Been Asked If They Want to Switch to EDI?   A surprising number of companies prepare themselves for EDI transactions without ever asking trading partners if they want to participate! Contacts with trading partners should commence as soon as the EDI project starts, so that they develop EDI systems in concert with the company.

What Is the Ramp-Up Schedule?   The cost of the EDI installation (including design, purchased software, hardware, and education) will take a long time to recoup unless many partners representing a large volume of transactions are brought onto the system quickly. The key to a cost-effective implementation is the speed with which trading partners are switched to EDI.

Will a Sales Program Be Developed to Educate Trading Partners in the Benefits of Using EDI?   Many trading partners are unwilling to switch to EDI and require some persuading. Some VANs provide marketing services that can help in this area.

Has a Startup Package Been Assembled for Bringing in Each Trading Partner?   Each time another trading partner starts to use EDI, there are several steps to complete before bringing the partner into your EDI system. First, review a legal agreement regarding the use of EDI. Then, define the types of EDI transaction sets that will be sent and received, so that both parties can modify their translation software. Also check if each trading partner's VAN will interchange transaction sets with the other partner's VAN. Finally, send each other sample EDI transaction sets, and be sure to include all possible data variations in the test, so that the ability of each company's translation software to process the information can be tested.

Should EDI Transactions Occur in Batch or Individual Mode?   EDI transaction sets can be sent in a cluster or individually. If communications links will be directly with trading partners instead of through a VAN, then individual transmissions are more likely.

How Will the Incoming EDI Data Be Validated?   The data in incoming transaction sets need to be reviewed for accuracy of such information as product codes, dates, customer codes, and quantities. This is a complex area, and varies by transaction set.

Will Interchange Control Numbers Be Used?   The software should be able to number each exchanged document in sequence, so that missing or duplicate documents are flagged; when the system misses a number or receives two of a number from a trading partner, it should notify the operator with a message.

Will Translation Software Be Developed or Bought?   Translation software is available (most VANs can recommend software packages), but some companies with in-house expertise may elect to develop the software themselves.

Should Data Go Straight to Customers/Suppliers, or Go Through a VAN? If only a small number of trading partners will be using EDI, then direct access to them may be the cheapest route. In most cases, however, it is easier to use a VAN, thereby avoiding problems with communications protocols and line speeds, busy phone lines, and restricted messaging hours.

Are Data Communications Software and Hardware Available?   Is the communications software already in-house? If direct communications with trading partners is contemplated, then equipment may be required for multiple systems.

Should Transaction Sets Be Sent Automatically?   The software should be able to automatically call the VAN at least once a day for messages without any operator intervention.

What Are the Error Recovery Contingency Plans?   When transactions sets are sent but not received by the intended party, or received in a garbled manner, the original transaction should be available for resending. Also, a time-out period should be built into the system so users will be warned if a response to a transaction set has not been received within a specific time frame and follow-up actions can be taken.

What Security Is in Place?   This topic is handled in some detail in the following section.

Are Expectations for Replacing Manual Systems Realistic?   The project team must realize that the company will never convert to 100% EDI transactions. There will always be small suppliers who will not use EDI because the number of transactions will be too low. Therefore, some manual processing must be maintained.

If a Pilot Project Is Used, What Criteria Will Define a Successful Pilot Test?   A pilot project should include sending and receiving all possible transaction sets as well as using a full range of data within each transaction set. Transactions should be 100% successful before a pilot project is declared acceptable.

SECURITY

The traditional control systems are all related to protecting information that is passed among employees on paper. For EDI, control systems must be completely redesigned. Some of the security issues involving EDI are as follows.

Auditing the Transactions

If there is no paper trail, how are EDI transactions audited? The choices are to rely on transaction records kept by trading partners, by the VAN, or internally. If a trading partner is relied on for paper records, the auditor must be concerned that the trading partner is also trying to cut back on paper records (that it is using EDI) and therefore may not have any paper records either. A VAN is possible, but the number of EDI transactions that flow through a VAN in one day would create a remarkable amount of paper documents (and would require additional fees to have the VAN create and store the records). The best alternative is to rely on internal records to satisfy the needs of the auditors. The company may either keep paper records of all transactions (which circumvents the EDI objective of eliminating paper) or work with the internal and external auditors to implement adequate controls while the EDI system is initially being designed and implemented. The key is to involve auditors while the system is being designed, so that changes to the control systems are relatively inexpensive. Changes to these after the EDI system is operational are much more expensive to implement, since the controls must be designed into preexisting computer code that must be modified to accommodate the changes.

Eliminating Authorization Signatures

EDI transactions do not include signatures. Instead, personnel issuing transactions through an EDI system are preauthorized to issue transactions up to certain dollar commitment levels. Transactions above those levels must be authorized by a supervisor. A way to circumvent this system is for an employee with a low level of authorization to send a large number of small-dollar EDI transactions (easy to do, since the system is automated). This type of malfeasance can be found after the fact by an auditor review or by an automated review of unusual numbers of small-dollar purchase orders, using statistical analysis techniques.

Verifying Transmitted Data

Many new EDI users are concerned that EDI transactions will be lost by the recipient's computer system. This problem is solved by the acknowledgment messages that are sent back to the sender by the recipient. Unfortunately, this only means that the complete message was received; it does not mean that the recipient's computer, production, and shipping systems are capable of processing the order. This is an issue that exists irrespective of the use of EDI as the method of transmitting transactions.

EDI TRANSACTION SETS

The more common transaction sets involve price catalogs, requests for quotes and related responses, purchase orders and related changes and acknowledgments, inventory inquiries, shipment notifications, invoices, and bills of lading. An example of the many types of transaction sets that can be used between trading partners is shown in Figure 8.6. Several hundred transaction sets are in use or under development. Most companies will only use a fraction of the total number of transaction sets, since many of them were designed for specific industries and are of no use in other industries. A sampling of some transactions follows.

- Advance shipping notice. Sent from the supplier to the customer, describing the quantities and types of materials in an outgoing shipment as well as the approximate arrival date at the customer warehouse.
- Arrival notice. Sent from an ocean carrier to a land carrier as well as the customer, notifying of the arrival of a shipment.
- Booking confirmation. Sent from an ocean carrier to a shipper, confirming acceptance of a freight booking for ocean transport.
- Booking request. Sent from a shipper to an ocean carrier, asking to place freight on a ship; includes any special handling information.
- Customs declaration and customs release. Used to process a shipment through customs.
- Financial information reporting. Sent from a bank to its customer, noting report balances and financial transactions.

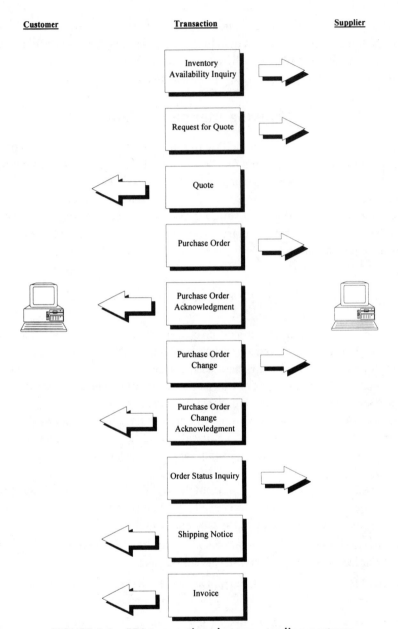

FIGURE 8.6   EDI transactions between trading partners.

- Freight details and invoice. Sent from the carrier to the supplier, detailing information about the carrying of freight and its related cost.

- Gate activity. Sent by the terminal operator to the ocean carrier to let it know that the cargo has been removed from the ship and passed to a land carrier for delivery to the ultimate customer.

- Inventory availability inquiry. Sent from a customer to a supplier, asking for information about available stocks of materials.

- Invoice. Sent from the supplier to the customer; itemizes the items being billed.

- Material release. Sent from the customer to the supplier; notifies the supplier of the need for more materials to be shipped to the customer.

- Lockbox transaction. Sent from the bank to the supplier; reports remittance information.

- Order status inquiry. Sent from the customer to the supplier; requests information about the status of an order.

- Payment cancellation request. Sent from the customer to a bank, canceling a previous EDI transaction and requesting that money be sent to a supplier.

- Payment remittance. Sent from the customer to the supplier; identifies what line items on the invoice are being paid. Can also be sent to a bank, authorizing payment to the bank of the supplier.

- Planning schedule. Sent from the customer to the supplier; contains information about forecasted manufacturing requirements.

- Price sales catalog data. Usually sent by the supplier to the customer; contains information to update an electronic catalog.

- Product information transaction. Can be sent by either the supplier or the customer; contains updates to information about a product.

- Purchase order. Sent from the customer to the supplier, requesting delivery of materials or services.

- Purchase order acknowledgment. Sent from the supplier to the customer; acknowledges receipt of the customer's purchase order as well as product availability and other details.

- Purchase order change. Sent from the customer to the supplier; notes any changes to the original purchase order.

- Purchase order change acknowledgment. Sent from the supplier to the customer, acknowledging receipt of a change to the purchase order and detailing any resulting changes to the price and delivery date.

- Purchasing transaction. Sent from the customer to the supplier, requesting delivery of a product or service.

- Quote. Sent by the supplier to the customer, responding to a request for quote. The quote contains information about price and delivery dates.

- Receiving advice. Sent by the recipient to the sender of a shipment, commenting on the condition and quantity of the shipment.

- Request for quote. Sent by a customer to a supplier, requesting a price for a specified product or service.

- Shipment information. Sent by a carrier to either the sender or the recipient of a shipment, in response to a request for shipment status.

- Shipment inquiry. Sent by either the sender or the recipient of a produce shipment, inquiring about the status of the shipment.

- Statement. Sent by a supplier to the customer; lists invoices sent during the previous period.

- Status details reply. Sent by a carrier; response to a shipment inquiry.

- Stow plan. Contains information regarding the location on an ocean carrier where a shipment is stored.

- U.S. customs manifest. Sent by an ocean carrier to the U.S. Customs Office; contains information about the contents of the ship.

EXPANDING THE EDI PROGRAM

A typical EDI installation is initiated to realize benefits from trading with a relatively small number of trading partners. These are the suppliers or customers who account for a large part of the company's sales or purchases. Their transactions with the company are both frequent and large. After these trading partners are included in the EDI program, many EDI projects languish without drawing in any additional trading partners. The effort required to bring on each additional trading partner is considerable, and the benefits gained from each conversion become increasingly small as the company converts progressively smaller trading partners to EDI.

The cost of gaining additional EDI partners should be analyzed in terms of the time period required to earn back the up-front fixed cost of adding a new partner with savings on cheaper transactions. Each additional partner requires time to educate it on the benefits of EDI and possibly travel costs to send people to the trading partner to assist in setting up the installation. However, once the initial expensive programming costs to convert and link incoming EDI transaction sets to the company's applications programs have been paid for, the cost of adding additional EDI partners is less than might be expected. Also, the cost may be further reduced by using the marketing assistance of a VAN. It is in the interests of a VAN to add trading partners to its network, since it makes money from the volume of EDI transactions going through its computers—and the more EDI trading partners, the higher the volume. Thus, some VANs may be willing to assist in educating prospective trading partners about the benefits of installing an EDI system.

COST/BENEFIT ANALYSIS

When an EDI installation is first proposed, it should be accompanied by a cost/benefit study that itemizes the specific cost and savings associated with the project. To provide a template for constructing a company-specific cost/benefit analysis, the following example is included. This example is only meant to be an approximate template. Cost and benefit information should be generated in much greater detail for a real analysis, including a discounted cash flow analysis over several years.

Install an EDI System

Mr. Sanderson has bought Patio Units Deluxe (PUD). One of his first decisions is to tie his automated order entry system to the payment system of PUD's largest customer, Walters Outdoor Equipment Stores (WOES). Currently, 65% of the company's business is with WOES. The revenue of PUD last year was $11 million. The average order from WOES for a patio unit shipment is $500, with (on average) four payments clustered into each check, for an average check amount of $2,000. It takes 10 minutes to enter a check payment into the company's cash application system.

The cost of a PUD accounting clerk is $11 per hour, which includes the cost of benefits. The company's auditors issued a letter to management along with last year's financial audit that stated the company had lost $15,000 last year because of errors made while entering cash receipts into the computer system. The cost of purchasing EDI software is $2,500. The easiest way to set up the EDI system is to use a value-added network (VAN) to transfer EDI transactions from WOES. The charge per VAN transaction is 40 cents. The PUD programmer estimates that writing the software for the interface between the EDI software and PUD's accounting software will take three months of her time. The programmer earns $33,000 annually. Should the company undertake this project?

Solution.   The cost of implementing EDI is balanced against the savings from eliminating losses due to incorrect entry of cash receipts and from eliminating the labor to enter payments into the accounting database.

Benefit of Implementing EDI

Eliminate losses from errors	$ 15,000
Amount of WOES business	$7,150,000
Average payment size	÷ $2,000
No. of WOES payments/yr	3,575
Time to enter one receipt	× .17 hr
Time/yr to enter receipts	608 hrs
Cost of receipts clerk/hour	× $11
Total labor savings	$ 6,685
Total saving/yr	$ 21,685

Cost of EDI Implementation

Cost/yr of programmer	$33,000
Three months' work	× .25
Total programming cost	$ 8,250
Cost of packaged software	$ 2,500
Total software cost	$10,750
No. of WOES payments/yr	3,575
Cost/VAN transaction	× $.40
Total cost/yr of transactions	$ 1,430
Total implementation cost	$12,180

Since the total cost is $12,180 and the total benefit is $21,685, the cost/benefit analysis indicates that an EDI system should be implemented.

EFFECT ON PAYMENT SCHEDULES

EDI does not necessarily change the speed with which a company pays its bills. Several centuries ago, credit was extended for up to two years on individual business transactions because the transactions (such as sailing to the Pacific for whaling, or around Africa to barter for goods in India) took that long to complete. Now, with rapid transaction completion with the assistance of such techniques as EDI and JIT, the reason for prolonging the payment period is not so clear.

However, many EDI users continue to use a 30-day payment schedule, while sometimes offering early-payment discounts. The reason for these payment terms is partly related to negotiation; a company can earn money on cash it is not required to pay to a supplier, so the company makes money by prolonging payment. Also, though the initial transaction of ordering and receiving goods from a supplier is short, the time needed by the purchaser of the goods to resell them may be quite a bit longer. If the purchaser's selling cycle is long, then the purchaser may be unwilling to pay for the goods until the selling cycle is completed and the cash has been received. Thus, even though the technology exists to pay a supplier immediately, other business reasons may act to prolong the payment of bills.

EFFECT ON FINANCIAL RATIOS

The advent of EDI may have an effect on the acceptability of certain ratios. If a company were to use EDI for all transactions, the turnover of inventories, receivables, and payables would greatly accelerate, resulting in minimal amounts of all three items on the balance sheet (this assumes that receivables would be paid immediately with electronic funds transfers, which is not yet a common practice even among heavy EDI users). As a result, the following ratios that include inventories, receivables, or payables would be highly skewed.

Ratio	Direction of Skew
Current ratio	Lower
Quick ratio	Lower
Ratio of net sales to receivables	Higher
Average collection period	Lower
Turnover of inventories	Higher
Turnover of current assets	Higher
Ratio of net sales to working capital	Higher

The changes imposed by EDI may render useless any comparisons of companies within industries, since ratio results will vary so widely between EDI and non-EDI firms. Only industries where all major companies have converted to EDI will show comparable ratios.

TERMINOLOGY

The following terms are frequently used in relation to EDI systems and may be useful to the reader in researching this topic in more detail.

- Applications link. The software that reformats arriving EDI data into a format compatible with a company's applications programs. Alternatively, it may translate data from a company's applications programs into an EDI-compatible format for transmission purposes.

- DUNS number. Created by Dun & Bradstreet to uniquely identify a company; used by many trading partners to identify themselves in EDI transaction sets.
- Industry-specific standard. Some EDI standards are designed for specific industries, since they contain transactions not used by any other industries. For example, UCS is used by the grocery industry, and ORDERNET/NWDA is used by the pharmaceuticals industry.
- Payment from receipt. When a customer pays its supplier based on the quantity of product received. An invoice is not used.
- Smart card. A floppy disk that contains the invoice for a shipment. The disk is brought to the receiving company by the supplier when its materials are delivered. If the invoice is correct, it is uploaded to the receiving company's computer, or is corrected and uploaded. In either case, both trading partners now have identical copies of the invoice.
- Translation software. A program that maps the data between the data formats used by EDI software and applications software.
- Transaction set. The header, detail, and summary parts of an EDI transaction.
- Transmission acknowledgment. The transaction sent from the receiving party to the sending party of an EDI transaction, acknowledging receipt of a full transmission. The acknowledgment is sent prior to translation into a format readable by the recipient's applications software. This means that the received transaction has not been checked for validity when the transmission acknowledgment has been sent.
- Value-added network. A company that provides electronic mailboxes for transmission of EDI transactions between companies, as well as other services such as line speed conversions.
- Vendor-managed inventory. When a supplier replenishes stock for a retailer; usually requires rapid response.

SUMMARY

In summary, this chapter discusses the history of EDI, how it can be used most profitably in a company, and where the most common pitfalls are in a typical EDI implementation. Ancillary topics include key terminology used in the EDI arena, a list of organizations to contact for more information about installing EDI systems, and a cost/benefit study that can be used as a rough template for a real analysis. When combined, all these topics should give the reader enough information to make a decision about proceeding with the first exploratory steps of an EDI system implementation.

REFERENCES

The 1992 EDI Directory. Potomac, Md.: Phillips Publishing, 1992.

Barber, Norman F. *Organizational Aspects of EDI: A Project Manager's Guide.* Electronic Data Interchange Association.

Emmelhains, Margaret Ann. *EDI: A Total Manager's Guide.* New York: Van Nostrand Reinhold.

Kutten, L.J., Bernard D. Reams, Jr., and Allen E. Strahler. *Electronic Contracting Law, EDI and Business Transactions.* 1992–1993 ed. New York: Clark Boardman Callaghan.

Macht, Joshua. "Are You Ready for Electronic Partnering?" *Inc.,* November 14, 1995, pp. 43–51.

Powers, William J. *EDI Control and Audit Issues for Managers, Users and Auditors.*

Shaw, Jack, and Mike Witter. *The EDI Project Planner.* EDI Executive Publications,

Sokol, Phillis K. *From EDI to Electronic Commerce.* New York: McGraw-Hill, 1995.

Willson, James D., James P. Colford, and Janice M. Roehl-Anderson, *Controllership: 1995 Cumulative Supplement,* 302–313. New York: Wiley, 1995.

More information about EDI systems, case studies, transaction sets, and industry work groups can be obtained from the following organizations:

Secretariat
ASC X12 Data Interchange Standards Association, Inc.
1800 Diagonal Road—Suite 355
Alexandria, VA 22314
(703) 548-7005

Uniform Code Council, Inc.
8163 Old Yankee Road—Suite J
Dayton, OH 45458
(513) 435-3870

Electronic Data Interchange Association
1101 Seventh Street, N.W.
Washington, D.C. 20036-04775
(703) 838-8042

More information about EDI can also be obtained at the following trade shows, committees, and conferences: Data Interchange Standards Association (DISA) trade show; EDI Coordinators Conference; American National Standards Institute X12 committee meetings; Transportation Data Coordinating Committee (TDCC); Petroleum Industry Data Exchange (PIDX); Electronics Industry Data Exchange (EIDX); Electrical Industry Data Exchange Group; Chemical Industry Data Exchange Group (CIDX); and Automobile Industry Action Group (AIAG).

9

THE QUICK CLOSE

In most companies, information from the previous month should be available within the first few days after the end of the period. There is considerable variation among companies in the time required to periodically close the books. A 1994 study found that the average world-class company closed its books in four days versus seven days for an average company.[1] A prime determinant in the length of closings is the need to collect financial information from subsidiaries and merge it into the summarized corporate financial statement. Even with this addition to the length of the close, world-class companies out-perform average companies.

Computers can speed up the process of delivering information to management, but the processes being automated are frequently in need of overhaul as well. In fact, without a careful review of the underlying systems, new computer automation may speed up a process that is not even needed. This chapter discusses a number of process improvement steps to excise unnecessary items from the closing process and thereby make computer processing of the remaining steps more efficient.

High Quality is a key component of the fast close. Why should the controller consider quality in the closing process? Manufacturing operations have so far been the primary focus of quality improvement efforts, resulting in decreases in product costs, setup changeover times, and cycle times. Recently, the same improvement techniques have been used by controllers to improve their operations and control their headcounts. It has been estimated that efforts to improve quality can reduce the duration and cost of closing activities by 25% to 40%. That is the estimated amount of effort required by accountants to correct errors, eliminate roadblocks, and ensure the accuracy of their data. Many of the points in this chapter related to reducing the time required to close the period are essentially quality improvement recommendations.

---

[1] "Accounting Practices Benchmarking Study Spots Mistakes Companies Make," *Journal of Accountancy,* March 1995, p. 24.

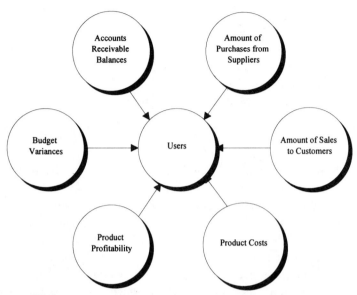

FIGURE 9.1   Accounting system user information needs.

Finally, this chapter discusses ways to reduce the time needed to close the accounting period. However, does the monthly accounting period need to be closed? For example, the financial information shared with company managers is so minimal that, from the perspective of the managers, there may be no need for anything more frequent than quarterly statements. In fact, a small number of companies have switched to quarterly closings, thereby cutting the annual amount of accounting work devoted to periodic closings. If a quarterly close is used, the controller can place selected monthly information needed by managers in the company's management information system, allowing managers to quickly access such information as sales by customer, outstanding receivables by customer, job cost status, profitability by product line, ongoing department costs versus budgets, and fixed assets assigned to various departments, as shown in Figure 9.1. Managers must be trained in database access techniques so that they can find the information. However, if the CEO requires company managers to be informed about company statistics on a frequent basis, they will have to access the information regularly.

REDUCING THE LENGTH OF THE CLOSE

A high-quality manufactured product is one that meets the customer's specifications and is delivered on time and at the right price. How can that definition be applied to the accounting closing process? To answer that question, we must look at each part of the definition in detail.

Who is the customer? The customer is whoever uses the output from the process. In this case, the customer is any user of the financial statement, usually lenders, investors, and managers.

What is the customer's product specification? The specification of this set of customers is to receive financial information that is accurate. The definition of the word *accurate* is crucial, for perfect accuracy is expensive in labor and time requirements. For example, precise cut-offs of accounts payable require extra time to wait for every supplier invoice related to the last period to arrive in the mail—why spend time waiting for a small invoice to arrive that has no major effect on the financial statements? However, if *accuracy* is defined as information that will not lead to incorrect decisions by the customer, then the controller can cut both the labor and time required to issue the financials by using selected estimates for them and adjusting the estimates after the period close. To use the same example, the controller can accrue costs based on items that have been received according to the receiving log, but for which an invoice has not yet been received from the supplier.

When is delivery required? The optimum delivery time should be at midnight on the last day of the accounting period. This may seem impossible, but the goal should be set. For example, if the current time requirement to close the books is 18 days, then try to shave a day off the process, and once that goal has been reached, continue to pursue an instantaneous close as an ongoing process.

What is the right price? Closing the period does not add any value to the product received by the company's final customer, so the effort going into it should be considered a non-value-added activity. Consequently, the goal should be to close the period at minimal cost, ideally using zero labor and assets. Again, this goal may seem impossible, but the ongoing process of reducing costs must be established. An opposing view is that the information supplied by the controller is invaluable and that all necessary costs should be incurred to supply high-quality information. The response to this view is that the accounting department is a cost center. It generates no profits, therefore its task is to produce its product (the financial statements) at the lowest possible cost to the company.

Thus, we now have a definition for a high-quality close of an accounting period: The product of the accounting close must provide information to lenders, investors, and managers that will not lead to incorrect decisions by those users. The information must be provided immediately after the close of the accounting period at minimal cost to the company.

The goals sound impossible. Let's look at ways to make them possible. The following set of sequential steps, when processed iteratively, will help the controller gradually reduce the time required to produce financial statements.

1. Clear out the junk.
2. Document the process.
3. Eliminate duplication.
4. Defer routine work.
5. Automate standard items.
6. Set investigation levels.
7. Move activities into the previous month.
8. Reduce cycle time.
9. Automate manual processes.
10. Replace serial activities with parallel activities.
11. Rearrange work space.
12. Train the staff.
13. Do it again.

Note that this sequence should be repeated continually, for additional processing improvements can always be found. In addition, new technological developments may appear that will reduce the processing effort even further, and these should be incorporated into the process as they are perfected. For example, using electronic data interchange (EDI) for receiving accounts payable will eventually allow the controller to close the accounts payable subsystem immediately following the end of the accounting period, with no time lag whatsoever. A fast closing requires a well-coordinated organization, teamwork, and good leadership. The controller must lead the process of continually reducing the time and effort required to achieve a fast closing.

Clear out the Junk

This step is similar to cleaning out your garage—you must throw out the trash so you can have a better look at what remains. As an example, Figure 9.2 shows two items that can be eliminated: one is an unnecessary restatement of the receivables aging report, and the other is multiple filings of copies of the same invoice. Possible solutions are to write the collection commentary directly onto the aging report (thereby eliminating the restated report) and to store the invoice on the computer for easy access (thereby eliminating all four invoices that were filed as well as a few filing cabinets). In general, the following points should be followed:

- Eliminate items that require multiple approvals; one should be enough.
- Eliminate items that must be filed multiple times (e.g., alphabetically, numerically, by state); once should be enough.

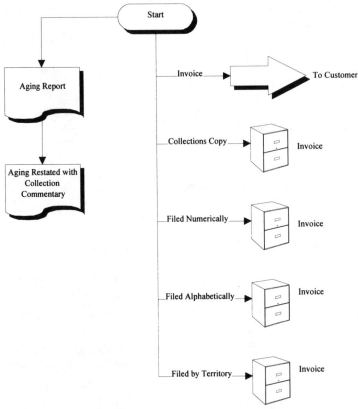

FIGURE 9.2   Examples of unnecessary receivables activities.

- Clean up the accounting area. If it is inundated with paper, then either file the paper away or (even better) review its usefulness and, if it is not useful, throw it away. This can include a clean-desk policy—at the end of the day, everyone's work must be filed away.

Document the Process

Do not implement solutions without first reviewing the process in detail. You should use both of the following techniques for documenting the process before continuing.

Create a Process Flowchart. This is a quick review of how the process flows. There are few symbols to remember; just start and stop indicators and a rectangle for each process step. The only other symbols needed are one for reports and a diamond for decisions points. The process should flow either top to bottom or left to right. Do not try to loop around the page, since the process can look too confusing. Instead, continue onto additional pages if the process description becomes too lengthy. After the process has been charted, go back and note the time required to perform each step, plus the queue time between steps. This information highlights the slowest points in the process. Finally, list the job title of the person who performs each step. This information highlights points in the process where move and queue times are likely, since these typically occur when a new person enters the process. A basic process flowchart is shown in Figure 9.3, with wait times and process times added in Figure 9.4.

The most immediate problem found on such a flowchart is the wait time as documents move between queues of various employees. Individual process steps are normally much faster than wait times. One solution is to reduce the number of people handling each step. This usually results in a small number of people handling a large number of process steps. The problem created by concentrating multiple steps with a small number of people is that they must be thoroughly trained in the added steps. Otherwise, insufficient training will lead to a high error rate, which will then require valuable management time to adjust.

Create a Geographic Flowchart. A geographic flowchart, like the one shown in Figure 9.5, shows where paper travels during a process and allows the controller to determine where wasteful movement occurs. In Figure 9.5, the flowchart indicates that the controller should swap the locations of the receivables clerk and the general ledger accountant in order to reduce the receivables clerk's move time to the copier room. This type of flowchart can also be used for tracking moves to fax machines and storage rooms.

Eliminate Duplication

Duplication typically occurs in two places during the closing process. First, information compiled at a subsidiary location is cross-checked at the consolidating location. Second, a subordinate's work is reviewed by a supervisor. Using the flowcharts developed during the preceding stage, highlight these duplications and eliminate them. Elimination is not easy if the cross-checking is done to fix numerous mistakes caused earlier in the process. The process must first be made error-proof. Then a further review is not necessary. Here are some examples of error-proofing steps.

Recurring Journal Entries Are Incorrect. Have an experienced staff person create entries for all periods of the year at one time and enter them all in the accounting software at once (however, some accounting packages do not allow this). Then the entries do not have to be recalculated each period, with

FIGURE 9.3   Basic process flowchart for closing the period books.

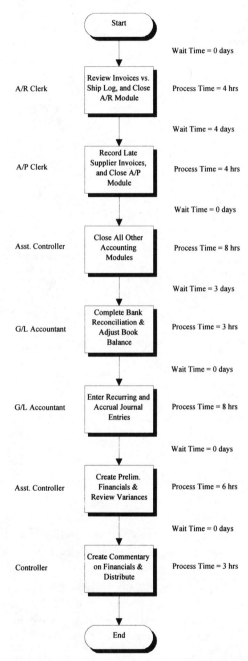

FIGURE 9.4 Process flowchart with functional responsibilities and process/wait times.

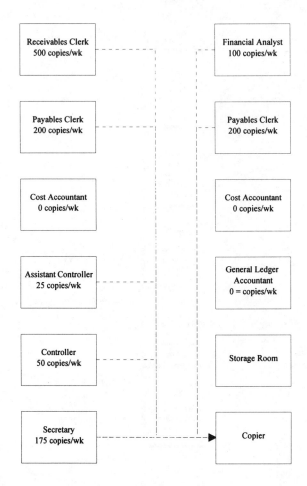

FIGURE 9.5   Geographic flow chart for an accounting department.

the possibility of errors occurring. If the recurring entries tend to vary slightly (e.g., when rent invoices are accompanied by a pro rata share of the building maintenance and utility fees), then enter the unchanged portion of the journal entry (e.g., the rent), thereby only leaving the variable portion (e.g., the maintenance and utility fees) for which an entry must be made. By splitting the fixed and variable portions of the invoice, the fixed portion can be entered for the entire year.

Items Shipped Are Not Billed.   Create a check-off column in the shipping log. The receivables clerk must note the invoice number and amount of the billing right in the shipping log. Then, if an item in the log has no invoice number next to it, the item should be investigated before being billed to the customer.

Items Received Are Not Invoiced by Supplier. Create a check-off column in the receiving log. For the last ten days of the period, the payables clerk must note the payables invoice number and amount of the charge right in the receiving log. Then, if an item in the log has no invoice number next to it, the clerk should accrue the cost based on the purchase order amount.

Costs Are Allocated to Wrong Account Numbers. This is when a budget column on the financial statement helps. Look for rows in the financials where the actual expense has no budgeted amount next to it at all—if there is no budget, the odds are high that the expense was charged to the wrong account. And, of course, if the variance between actual and budgeted costs is large, there may also be a mischarge. This comparison can be done at any time during the period, not just during the closing process. A quick look at the financials a week before the close will highlight these problems and leave time to correct them before the pressures of the closing process commence.

A great way to error-proof the process of charging to account numbers is to constantly review all accounts, and eliminate or consolidate those that are not being used. If the account numbers are no longer available, then no one can charge costs to them. To assist this step, many accounting software packages include a feature that keeps new activity from being charged to existing accounts.

Defer Routine Work

Take note of those items being performed during the closing process that are unrelated to it and that can be deferred until a later date. For example, performing account analysis on janitorial supplies is not crucial and can wait until the financial statements have been issued. Also, most accounts payable will not be paid for 30 days (except those with discount terms), so supplier invoices can be entered into the system at any time during the 30 days prior to payment (subject to payment approval procedures that vary by company).

Of course, the closing should not be used as an excuse to stop *all* other work. For example, invoices must be generated and mailed immediately after product is shipped or services have been completed, in order to maximize cash flow. In addition, payroll must be processed irrespective of the closing date.

Automate Standard Items

Prepare certain accounting entries on a standard basis and adjust periodically, as in the case of depreciation and insurance. This is a common area for errors, because standard entries eventually change, and the changes are frequently missed. To keep mistakes from occurring, the underlying documents that show change dates should be attached to the journal entry documents. For example, a property lease payment amount changes once a year. Rather than file the change schedule separately, either attach it to the journal entry or conspicuously note the change date on the journal entry form and cross-reference the source document. The change date can even be noted on the journal entry header in the computer record, so the general ledger accountant is prompted to make the change when the journal entry report for the current period is printed.

Set Investigation Levels

If variances are minor, their effect on the accuracy of the financials will be minimal. Investigation can safely be done after the financials have been issued, with any adjustments appearing in the financial statements for the following period.

The question is, What is the basis from which variances are calculated? The basis varies by account. For example, the revenue basis should be the total of the shipping log for the month. As another example, the accounts payable basis should be the receiving log total for the month (net of many other items, such as service costs and recurring payments such as lease and loan payments).

Another variance to be reviewed is any disparity between general ledger balances and subsidiary ledgers. The best guide to deciding what is an unimportant variance and what is worth immediate investigation is that as little as a 5% change in the company's net profit is considered to be misleading information. Thus, if the net effect of all variances that are not immediately investigated is a 5% change in net profits, then additional variances must be reviewed.

Move Activities into the Previous Month

Many tasks associated with the closing can be performed prior to the end of the period. Figure 9.6 shows examples of closing activities after the period-end that can be at least partly handled in the previous period.

Prepare Forms in Advance.   If journal entries or other reports are included in the closing process, then complete as much information as possible in advance, such as descriptions, account names, plan or budget data, and prior-period figures.

Anticipate Problems.   Be aware of areas where problems may develop and do as much analysis and reconciling as possible prior to closing. This could be the case where intercompany transactions are extensive and reconciliations may be difficult.

Develop Distributions.   Create as many allocation bases as possible before the end of the period. Allocation bases and certain ratios for distribution of costs may be determined in advance.

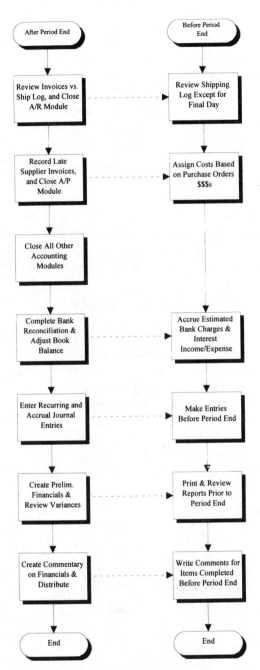

FIGURE 9.6   Closing activities that can be moved into the previous period.

Review Shipping Log.   The shipping log must be reviewed in its entirety to ensure that all items shipped were billed during the period. By reviewing the bulk of the log during the period, the period-end closing task is reduced to a quick review of the last day or two of shipments.

Accrue Estimated Bank Charges and Interest Income/Expense.   The bank can be periodically contacted by phone or modem to ensure that all special items, such as wire transfers, have been recorded on the company books. The interest expense or income can be estimated in advance, with an adjustment to actual being booked in the following period.

Enter Recurring Journal Entries.   Many journal entries, such as amortization, do not vary from period to period and can be entered into the general ledger in advance for the entire year (if the accounting software will allow it), thereby reducing the number of journal entries to be made during the period-end close.

Reduce Cycle Time

Cycle time is the total time required to complete a process. Frequently, the actual value-added time is minimal, whereas the wait time between steps constitutes the bulk of the process time. Target the longest wait times and take steps to reduce them. Figure 9.7 shows typical process and wait times for period closing activities, showing that the wait times needed for supplier invoices and bank statements

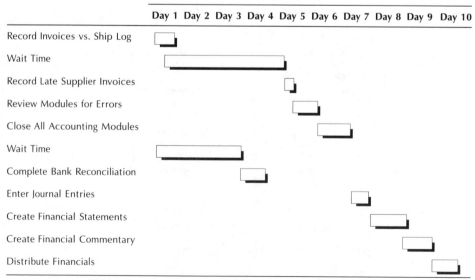

FIGURE 9.7   Process cycle times for closing the books.

to arrive are responsible for a large part of the time needed to close the books. Another common delay is the wait for closing information from subsidiaries.

Cycle time analysis also points out those process steps to which the controller can assign multiple staff, so that the steps can be completed more quickly. Typical steps that allow multiple staff assignments are accounts payable and accounts receivable. On the other hand, individual bank reconciliations are usually completed by one person.

Automate Manual Processes

After reviewing cycle times, select the manual processes that take large amounts of time and automate them. For example, taking inventory and extending unit costs to derive a total inventory valuation is extremely time-consuming. However, with an on-line perpetual inventory and automatic collection of unit costs through the accounts payable system, this activity can be reduced to a few keystrokes on a computer. Figure 9.8 compares the number of steps needed to close the inventory module if the inventory process is manual and if it is automated. The automated process has been reduced to a review for gross errors, whereas the manual system is clearly laborious. The following items related to the closing can be automated.

Cash.   Bank reconciliation software is now available for many accounting software packages. A typical system allows the user to call up a list of checks for the last period and mark them off in the system as having been cashed by the bank.

Inventory.   Perpetual inventory systems give the cost accountant reliable item quantity information for the end of the period. Note that truly accurate inventory quantities require continual vigilance by the warehouse staff and the internal audit team, which should conduct an inventory accuracy review *every week*. A second piece of automation is the costing of the inventory. Most accounting packages include a number of options for costing (e.g., FIFO, LIFO, average costing), where the accounting system automatically assigns costs to inventory items based on invoiced costs (for raw materials) or bills of materials (for work-in process and finished goods). The major problems with automatic costing of inventory are inaccurate bills of materials (corrected by frequent audits of the bills) and a lack of invoices at the end of the period for received raw materials (corrected by using the cost listed on the purchase order).

Fixed Assets.   Most fixed asset packages can reliably calculate depreciation. Some electronic spreadsheets do not stop depreciating when the accumulated

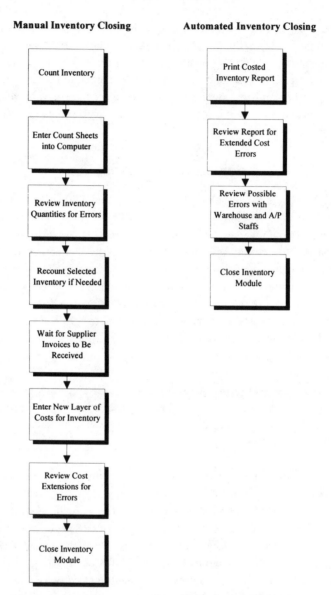

FIGURE 9.8   Manual versus automated inventory closing.

depreciation balance equals the total amount of the fixed asset, but an expert user of the spreadsheet can usually revise the depreciation calculation to avoid this problem.

Accounts Payable.   Some of the most advanced payables systems have eliminated the three-way matching process entirely. Waiting for late supplier invoices to arrive at period-end so that they can be matched to purchase orders and receiving records is one of the greatest drags on the closing process. The new payables technique of matching the purchase order to the received items right at the receiving dock, followed by payment based on the purchase order price, is a much faster way to close the payables module (see Chapter 4).

Replace Serial Activities with Parallel Activities

Based on the flowcharts developed earlier, identify steps that are currently performed in sequence but that could be performed in parallel, and then convert to parallel processing. For example, when the information on a report is needed by three people in order to generate their closing reports, don't wait for them to pass it along from one person to the next. Instead, give copies to all three so that they can process the information in parallel. Figure 9.9 shows four closing processes that can be completed in parallel, with minimal need for information sharing between the processes. The only need for serial activity is when the fixed asset processing must wait for late supplier invoices to be processed, in case a late invoice discloses the purchase of a fixed asset.

The major delay in most serial processes occurs at the beginning of the process, when there is a built-in delay caused by information not being available.

Waiting for Bank Statements to Arrive.   Bank statements typically arrive three days after the end of the period. To avoid this wait time, complete the bank reconciliation after the period has closed, and accrue any expenses related to the statement. Any adjustments can be booked in the following period. The major unrecorded expense on a bank statement is the interest income or expense. This information can be obtained by calling the bank (or accessing the account information directly by modem) and asking for the information. All other expense adjustments derived from the bank statement are typically small and have no effect on the income statement. On the other hand, missing balance sheet information such as unrecorded wire transfers can have a very noticeable effect on the balance sheet. To avoid this problem, review the cash book balance prior to month-end in comparison to the bank's records (on-line or by phone), and make any adjustments in advance. Then contact the bank immediately after the end of the period to confirm any unusual transactions during the final few days of the period. Not all transactions need to be reviewed (too laborious); instead, only review the transactions that tend to "slip between the cracks," such as wire transfers and voided checks.

Waiting for Late Invoices to Arrive.   Invoices relating to items received in the previous period can arrive five or more days after the end of the period and may still trickle in ten days after the period. To avoid this wait time, use

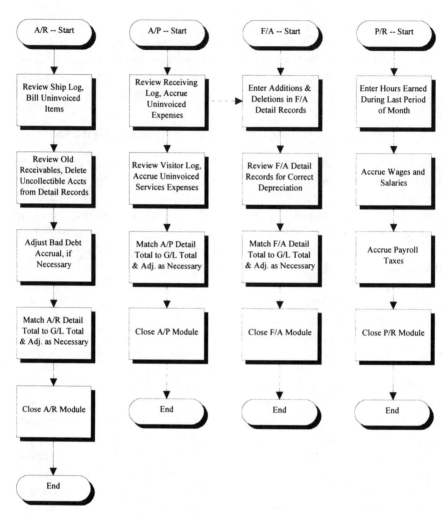

FIGURE 9.9   Examples of parallel activities.

purchase order records from the previous period. Any purchase orders from the previous period that have been recorded on the receiving log but not matched with an invoice should be accrued at the price listed on the purchase order. The journal entry can then be reversed for the next period, and the actual invoices recorded as they are received. A special problem is the accrual of service expenses that, by definition, do not appear in the receiving log (since they are services, they cannot be received). These expenses can still be accrued in two ways. First, many services are recurring, and contracts describing the periodic costs are on file. Simply review the contracts and record recurring journal entries for the amounts listed in the contracts. The second technique is both less accurate and more labor-intensive: review the visitor's log for the end of the last period, looking for sign-ins by service personnel (such as computer or building repair people). Then contact the companies represented by those people and ask for the invoice amount yet to be received.

Waiting for Subsidiary Closings to Occur.   Subsidiary closings usually take five to ten business days. There are numerous ways to avoid this wait time. For example, does the information need to be consolidated at all? If the financial information is only being used internally, management may be more interested in individual corporate operating results than in the summarized totals. If this is the case, then consolidations are only necessary for external reporting purposes, and the entities receiving that information may be content with longer wait times to receive their information. Also, very small subsidiaries have a minimal effect on overall corporate results, so preliminary financial information can be booked for consolidation purposes, and final adjustments can wait until the next reporting period. Another technique is to focus on the method of transmission to the corporate parent. The final subsidiary information can be transmitted directly into the corporate computer or sent by fax or overnight mail. A common approach is to transfer the information to an electronic spreadsheet template on a diskette, which is then overnighted to the corporate parent and quickly consolidated into an identical electronic spreadsheet.

Rearrange Work Space

After the closing process has been cleaned up and the paperflow made more efficient, the controller should review the work space.

Rearrange Work Area.   If the work area can be rearranged to reduce paper movement to a minimum and cut the level of nearby traffic (thereby reducing interruptions), then moving the staff is justified.

Use Cubicles.   Concentrating the staff in an open area with cubicles encourages more interaction among the staff. Too often, accounting personnel become ensconced in offices with locked doors because of a perception that all records must be kept close at hand in their offices and that the records are so confidential that the offices must therefore be locked. To avoid this problem, review all records and send them to a storage facility if they are not being accessed more frequently than once a month. Then concentrate the confidential files (usually contracts and personnel-related documents) in a minimal number of locked rooms. Finally, move the staff into more space-efficient cubicles, where the much-reduced files will now fit. Any overflow of documents can be filed in storage cabinets in an open space. Left-over offices can then be used as meeting rooms.

Use Low-End Copiers.   Most staff do not use the more advanced features of copier machines, such as stapling of multiple copies, double-sided copies, and large numbers of sorted copies (with the advent of electronic mail, long copy runs are much more infrequent, since a file can be sent to the entire company with a few keystrokes on the computer). Since a full-featured copy machine can occupy an entire room and be expensive to maintain, a cost-effective alternative is to purchase a large number of small copiers to be distributed throughout the accounting area, with just one full-featured machine available in a central location. This has the advantages of putting copiers closer to the staff, having many copiers still available in case one machine breaks down, and reducing the cost of copying. As an example of cost differentials, a mid-range copier with a full complement of features costs about $15,000–$20,000, whereas a slow copier with minimal features can cost as little as $300–$500.

Use a LAN Fax Gateway.   A PC can now be linked to a local area network's file server and set up as a fax machine for all users on the network. This allows PC users to send faxes through their workstations without ever leaving their desks. The fax gateway will even queue all faxes, redial busy phone numbers, and notify users via e-mail that their faxes have been sent.

Train the Staff

The accounting staff must be well trained in the closing procedures. Cross-training the staff will minimize the problems of peak loads, trouble spots, and absenteeism. The training should include indoctrination in the closing schedule:

- Tasks to be accomplished.
- Responsibilities for recording, preparing, analyzing, or transmitting information.
- Exact times (day and hour) by which each task must be completed.
- Specific cut-off dates by subsystem.
- Periods to be reported.
- Number of weeks to be included in each month or quarter.
- The day of the week on which the period closes.

Figure 9.10 shows a portion of a closing schedule. Reviews with all managers must be held before the schedule is finalized to ensure that everyone fully understands the closing process and can execute it. Any period-specific trouble spots (such as scheduled vacations) should be analyzed and solved at this time.

Task to accomplish: Close the fixed assets module.

Responsibility: Assistant Controller.

Timing: Must be closed by noon on the third day of the new period.

Notifications: Notify the general ledger accountant as soon as the module has been closed.

Specific steps to complete:

1. Review purchase orders for last five days of the period to ascertain the purchase of uninvoiced fixed assets.
2. Review previous period invoices to ascertain the purchase of fixed assets.
3. Record all new fixed assets in the detail ledger.
4. Review asset sale records for the period, and remove those items from the fixed assets detail ledger.
5. Record gains/losses on the sale of fixed assets during the period.
6. Review fixed asset depreciation amounts for depreciation in excess of the asset balance, and correct any excess depreciation amounts.
7. Compare the depreciation amount to previous-period depreciation amounts for reasonableness. Investigate any large variances.
8. Reconcile the general ledger fixed asset and accumulated depreciation balances to the detail ledger balances, and correct any variances.
9. Close the fixed assets module.

FIGURE 9.10   Portion of a closing schedule.

Do It Again

This improvement process must be repeated again and again in order to continually shrink the processing time. There are several reasons for this. First, bureaucracy tends to creep into a process via extra filing requirements, added steps, and additional approvals. Bureaucracy must be guarded against by constant review of the process. Second, the competition is always refining its processes. If competitors continue to cut their costs and cycle times while your company does not, then they will eventually gain an advantage through reduced overhead costs. Also, performance bonuses may be tied to continual improvements—ending the improvements may end the bonuses, leading to a drop in staff morale. If the process review is always ongoing, then employees will feel that they have a stake in the process, and they will perform better and support the continuing improvements.

There may come a time when the flood of improvements dries up. This is when the controller should look outside the organization for new ideas. The staff can be sent to seminars, college courses, or other companies to collect new ideas. Consultants can be brought in to review operations and offer suggestions. The sources of new ideas may be rather unusual; for example, flowcharting is more highly developed in the programming field than in most accounting operations.

Therefore, a computer consultant can be brought in or staff can be sent to a data-processing shop to visit or the accounting staff can work toward computer certifications or go to computer system flowcharting classes.

As another example, manufacturing operations employees are quite familiar with many techniques for reducing setup times, whereas accounting operations personnel have not yet considered the concept. Once again, manufacturing consultants can be brought in (several of the "Big Six" accounting firms have large groups of manufacturing consultants), or the accounting staff can review manufacturing materials written by such companies as the Oliver Wight Companies or the JIT Institute, which specialize in streamlining manufacturing processes.

In addition, interview your internal customers periodically to see if portions of their needed information can be sent to them prior to the issuance of completed financials, if there is additional information they need, or if there is information they no longer need. Finally, read the literature for ideas—there are many publications that help controllers who are interested in improving the quality of their closing procedures.

MISCELLANEOUS ISSUES

The following items should be considered by the controller when devising the closing process, since they have an effect on the speed of the close.

Create a Chart of Accounts

Develop a practical and uniform chart of accounts. This should include a proper grouping of accounts to ensure uniformity in reporting, both between segments of the business and from period to period. A typical account structure is as follows:

- Two digits for company codes (e.g., Widget Company or Plantagenet Company).
- Two digits for department codes (e.g., production, marketing, and engineering).
- Four digits for expense codes (e.g., rent, utilities, and office supplies).
- Hyphens to separate the codes.

Of course, single companies can omit the company code. The total account code should be kept short in order to avoid miskeyed numbers. In addition, the department and expense codes should be uniform across companies to assist in consolidations. For example, if Company A has the marketing department in code 920 and Company B lists the department as code 818, then an automated consolidation based on department codes can be difficult.

Set up Report Due Dates

Establish a firm schedule of management report due dates from which to determine the cut-off of various types of transactions and recording of accruals if applicable. Be sure to review the report list periodically to see if some reports can be deleted or modified, thereby eliminating the associated report preparation work.

Control Output

Control the release of information so that premature or incomplete data are not given to management. Otherwise, an excessive amount of time may be required to correct information that may have been inaccurate (e.g., recalling all issued reports and redistributing a new set of information). Also, management may make decisions based on inaccurate information. Finally, if the information has been sent to stockholders, bankers, or senior management, the reputation of the accounting department will suffer.

Use Exception Reporting

Where possible, use exception reporting to save management time. Certain variance levels can be reported, below which there is no effect and no reporting is needed. One possibility is to have the controller maintain a book that contains all company statistics, ranging from financial to operating ratios. This information is accessible on demand, but only large variations are reported to management each period. Another option is to report all statistics but to highlight large variations in bold font.

Create Branch Schedules

Realistic cut-off dates must be established for branches, overseas operations, and remote locations. It is very important that the corporate controller has clearly assigned the accounting responsibility for every operation of the business, thereby ensuring that there are no gaps in recording and no duplications. Use modems to transfer information directly to the headquarters computer to avoid data rekeying. The data format is important, since the corporate computer may not be able to read the transmitted information. A good low-tech approach is to fax the information to corporate headquarters or to deliver the information on a diskette by overnight mail.

To save time with the consolidated closing, very small subsidiaries can transmit expected period-end results almost immediately after the period end, with final adjustments arriving later, and being posted to the books for the following period. The risk of financial misstatement is small, since even the largest adjustment for very small subsidiaries should have no material effect on the consolidated financial statements. There are notable exceptions to this option, such as those companies dealing with derivatives; even a small company can then be responsible for large profit and loss swings—using preliminary reporting from such companies for the period close may backfire.

SUMMARY

In summary, a variety of techniques are available that allow the controller to close an accounting period much more quickly than before. For example, activities can be conducted in parallel instead of serial order, which shrinks the total closing time. Another method is to move activities into the period prior to the end of the month in order to reduce the number of activities that must be completed between the end of the month and the issuance of the financial statements. In addition, a number of methods are available to automate steps that are currently conducted manually; this reduces the amount of time required to complete the steps, and reduces the chance of errors creeping into the financial information. Taken together, these changes result in a considerable reduction of the time needed to complete the monthly financial statements.

REFERENCES

Allen, Kenneth A. "The One-day Close." *Management Accounting,* November 1995, p. 18.

Coburn, Steve, Hugh Grove, and Cynthia Fukami. "Benchmarking with ABCM." *Management Accounting,* January 1995, pp. 56–60.

Enzweiler, Albert. "Improving the Financial Reporting Process." *Management Accounting,* February 1995, pp. 40–43.

Hackett Group. *AICPA/THG Benchmark Study: Results Update and Analysis.* 1994.

Willson, James D., Janice M. Roehl-Anderson, and Steven Bragg. *Controllership.* 5th ed. New York: Wiley, 1995.

Zarowin, Stanley. "Motorola's Financial Closings: 12 "Nonevents" a Year." *Journal of Accountancy,* November 1995, pp. 59–63.

10

PROCESS DOCUMENTATION

An important tool for achieving revised accounting systems is process documentation. It is necessary to thoroughly document the flow of a process before it can be streamlined. After streamlining has occurred, the new process must be carefully documented so that the people performing the function can understand how to do it. Without proper understanding, mistakes will occur in the process, leading to considerable costs to correct the mistakes.

This chapter explains how to document both existing and prospective processes, using flowcharts and written procedural documentation that are combined into a procedures manual. The procedure being documented in this chapter as an example is the order-taking process. A manual should describe not only how to do a job but also who should do it. Therefore, a set of sample job descriptions is included. Figure 10.1 shows how a procedures manual is organized.

DOCUMENTING EXISTING PROCESSES

Documenting an existing process is not as simple as listing the sequence of events in the process. If we were reviewing an order-taking process in such a manner, we might derive the flowchart in Figure 10.2. Note the minimal number of flowchart signs needed. An oblong shows a process, and a diamond shows a decision point. Start and end points are also indicated. Many other flowchart signs exist, but these two signs can be used to document most processes.

This initial flowchart shows the processing steps but leaves out several key items:

- Time required to complete each step. Processing times must be documented, so that process improvement work will be directed toward shrinking the processes that take a long time rather than the short-time processes.

CONTENTS

FIGURE 10.1   Procedures manual table of contents.

- Wait time between processing steps.   Usually longer than the actual processing time, the wait times must be documented to highlight areas requiring improvement.

- Forms used as inputs.   Forms need to be documented in order to highlight those with duplicate information, or unused or missing information.

- Reports used as outputs.   Frequently, more reports are being generated than anyone suspects, because reports were requested for short-term needs and never canceled once the needs were fulfilled. Also, the information on several reports may be merged onto fewer reports.

- Who performs each step?   Wait times are often so long because the process steps jump between employees, and the wait time occurs because the item being processed enters each person's work queue and waits there while previous work-in-process is completed. A typical remedy is to concentrate consecutive task steps with the smallest number of employees, thereby reducing the number of work queues.

- Informal communication paths.   When a person needs information that is not being supplied by the formal system, the employee will go outside the system to locate the necessary information. By documenting these informal communication paths, they can be added into the new formal process.

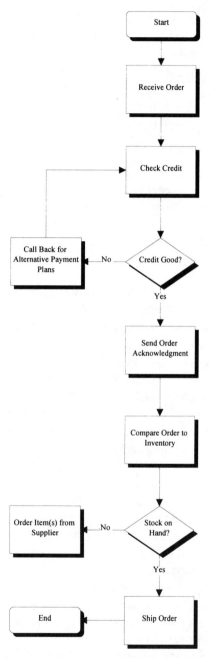

FIGURE 10.2   Order-taking process: Procedural steps only.

- Key control points.   Control points must be documented, so that changes can be made to the system with full knowledge of which controls will be affected by the changes.

- Errors.   It is important to find out where errors occur in the process, since error correction is very time-consuming. To do this, the average number of errors per thousand can be listed for each process step. An internal audit study of the process will be required to collect this error information.

Figure 10.3 shows the same flowchart but includes process and wait times. This information allows the user to prioritize activities that should be revised in the new process. It is clear from the figure that wait times are the primary culprit in causing long processing times. In particular, the wait time before comparing the order to inventory and the wait time before shipping the order keep a typical order from being processed in just a few hours.

There is little reason to spend time and money in enhancing the Receive Order step, since it only consumes 10 minutes. However, other considerations may require improvement of otherwise acceptable steps. For example, the marketing department may want to use "caller identification" in order to call up existing customer records prior to answering a phone call. This need would only slightly speed up the process but could be entered in the accompanying documentation of prospective changes and incorporate into the finished process.

Figure 10.4 shows the same flowchart but includes the forms used as input and the reports used as output. It shows an order form, which could be replaced by on-line entry directly into the computer, and a cumbersome process of creating individual purchase orders for each item requiring restocking, when in-quantity ordering may be simpler.

Figure 10.5 shows the same flowchart but includes the position or department performing each step. It shows when the process switches between people and enters the work queue of a new person. The process has a number of hand-offs between employees, resulting in long wait times between process steps. Based on this information, consolidating process steps with fewer people should eliminate a portion of the wait time.

Figure 10.6 shows the informal communication paths taken by employees to perform their assigned tasks. The amount of informal communication hints at inaccurate information in the corporate database. Since informal communications are being used to verify computer information in the areas of credit and inventory, it appears that the employees do not trust the information they are receiving from the system and are verifying it through informal networks. Obviously, audits of receivables and inventory accuracy are needed to find the underlying problems causing the inaccurate information.

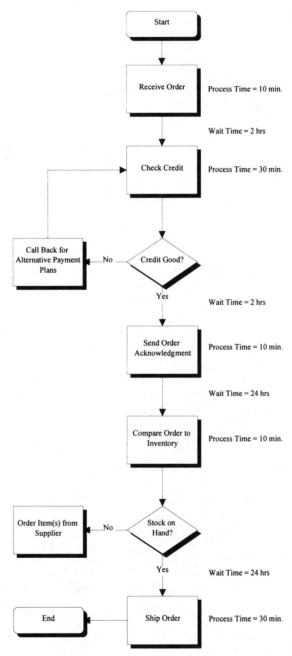

FIGURE 10.3   Order-taking process with process and wait times.

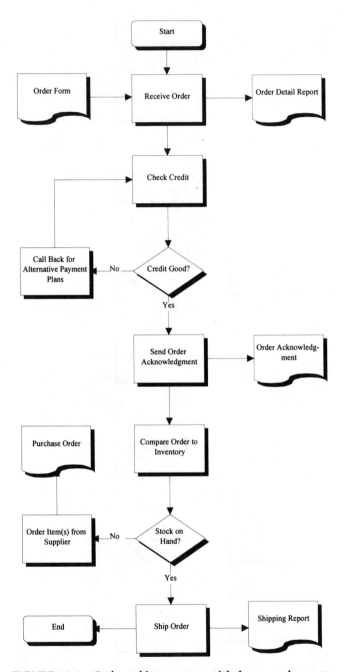

FIGURE 10.4 Order-taking process with forms and reports.

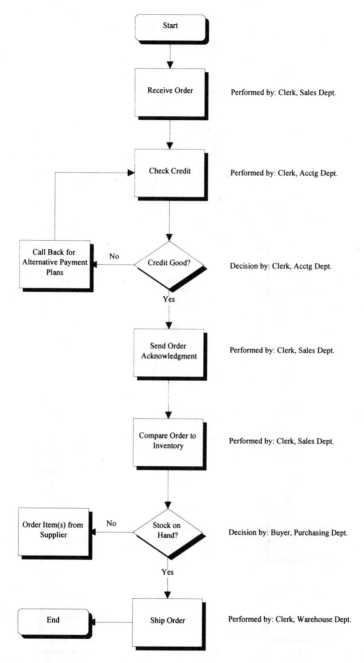

FIGURE 10.5 Order-taking process with responsibility for each step.

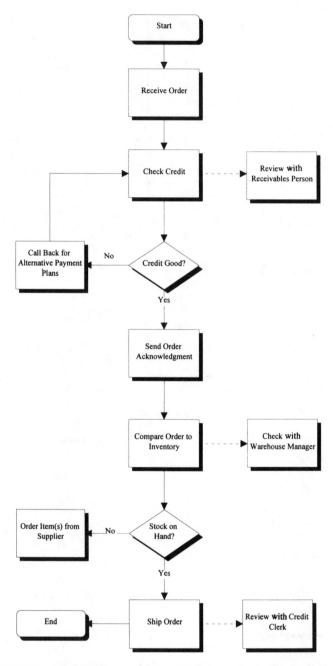

FIGURE 10.6 Order-taking process with informal communication paths.

Figure 10.7 shows only those interdepartmental controls related to not losing orders in the system. The presence of so many cross-checks indicates that loss of orders has been a problem in the past.

Figure 10.8 shows the average number of errors per thousand for each step of the order-taking process as well as a break-down of the specific error types. The data indicate a considerable problem with entering accurate information into the beginning of the process, with the effect cascading through the entire process. For example, errors are occurring in entering accurate customer addresses during the order-taking step, resulting in products being shipped to incorrect addresses at the end of the process. This problem has many solutions, such as on-line editing as orders are entered into the computer (e.g., do zip codes match the entered city name?), reading back item descriptions to customers over the phone, and auditing inventory accuracy levels, with follow-up on recommended inventory-tracking suggestions.

DOCUMENTING PROSPECTIVE PROCESSES: FLOWCHARTS

After the existing process has been reviewed, it is decided to make the following changes to the order-taking process:

- Merge the order-taking and credit review staffs, so that one person can handle both tasks. This will eliminate the 2-hour wait time while transactions flow between departments.

- Give the order-taking/credit review clerk read-only access to on-line inventory records, thereby eliminating the 24-hour wait time for information to come from the warehouse (this improvement assumes a computerized perpetual inventory system).

- Automatically create a purchase order for items not in stock, to be reviewed by a buyer before transmission to a supplier by mail or EDI. This eliminates the possibility of out-of-stock items not being reordered and speeds paperflow.

- Initiate periodic audits of inventory accuracy, leading to higher accuracy levels. This will eliminate the need for informal communications to cross-check the accuracy of the inventory records and will decrease the number of errors in the Compare Orders to Inventory step.

- Initiate periodic audits of the accuracy of customer credit history. This will eliminate the need for informal communications to cross-checks the accuracy of the credit records.

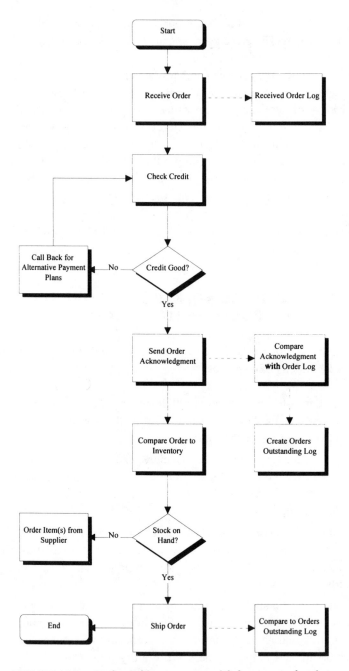

FIGURE 10.7   Order-taking process with key control points.

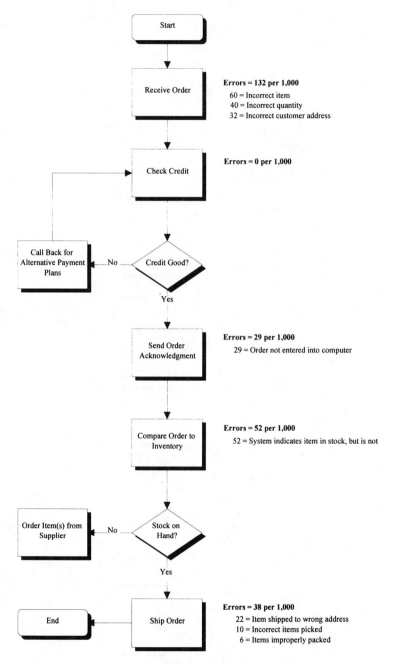

FIGURE 10.8   Order-taking process with errors noted.

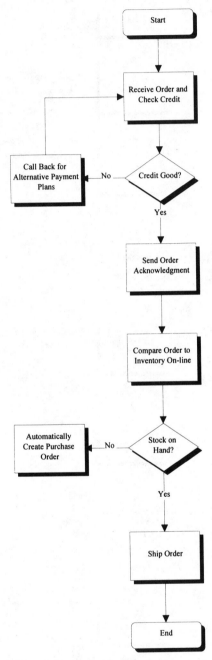

FIGURE 10.9   Revised order-taking process.

- Automatically compare records of orders received to orders shipped, and periodically audit the remaining orders to ensure that no orders are being lost. This will reduce the need for continual cross-checks at every step of the order-taking process.

The flowcharts shown earlier can then be revised to show the new sequence of events. However, those who use the process do not need to see all the flowcharts—they just need the basic process flowchart, so that they know how to do their jobs. The main point here is to briefly show the flow of the process, not to burden the user with an excessive number of details, like queue times and control points. The basic process flow of the prospective process is shown in Figure 10.9. Once the user has a summary flowchart that shows how to process an order, more detailed information must be added, so that all *common* process variations are covered. This information is written instead of being charted and is called either a policy or a procedure. Policies, procedures, and flowcharts are grouped together into a manual.

DOCUMENTING NEW PROCESSES: POLICIES AND PROCEDURES

Manuals contain policies, procedures, and flowcharts. A *policy* is a statement by top management to be used as a general guide by the organization on how to deal with a situation. A *procedure* is a specific rule or series of steps to follow in accomplishing a task. Procedures can be developed at any level in the organization. It is when policies and procedures are written down that manuals come into being.

All entries have policies and procedures. However, many have only informal systems whereby policies and procedures are created and modified daily; these systems result in inconsistent treatment of similar activities with resulting inefficiency and loss of uniformity. It is therefore important that the policies and procedures be written down and published.

This information will help supervisory employees in the day-to-day handling of their jobs, for the manual is an instrument of communication. It provides a common understanding of policy interpretations and clearly states the steps to be performed to complete tasks.

Purpose and Use of Manuals

As a business expands, responsibility and authority are delegated. Manuals then become more important, since they become the primary means by which management communicates its goals and controls the implementation of those goals. Also, manuals bind together dispersed but similar operations. They are the device by which the company says, "This is how we do it."

Policy and procedures manuals are normally prepared for each of the functional units, such as engineering, personnel, materials, quality control, finance, and accounting. Also, as matrix organizations have become more popular, manuals have been prepared for product lines that involve the activities of many functional areas.

Companies may create manuals to fulfill any of the following needs:

- To communicate to all management levels and satellite organizations the basic policies and procedures of the company.
- To provide a common understanding of policy interpretations and to define and clarify policy or procedures issues that may arise.
- To promote standardization and simplification: in the preparation of a manual, alternative courses of action or methods are considered and the most appropriate ones adopted.
- To train employees assigned to new jobs.
- To implement new procedures quickly and with greater understanding throughout the company.
- To allow timely communication of changes in policies and procedures.
- To create an internal control tool for management and form the basis for an audit of performance for compliance.
- To avoid duplication of effort by clearly stating responsibilities: a manual usually includes complete job descriptions for key positions.
- To reduce the amount of management time required to provide instructions and directions.

Organization of Manuals

The procedure-writing task may be assigned to either the department primarily concerned or to a departmental unit organized and staffed primarily for such a purpose.

The size of the company will probably dictate where the responsibility for procedure writing is placed. In smaller companies this function may be carried out by an individual under the direction of the controller. In larger companies the work may be a function of staff reporting to the president or general manager. There are several reasons for the controller to be involved regardless of where the responsibility is placed. Internal control is an integral part of the management process, and trained accounting personnel are uniquely aware of these requirements. Also, the controller's organization normally is involved in transactions that cut across departmental lines, and an overall rather than departmental approach to procedures is necessary. Many interdepartmental procedures involve paperflow that is of concern to the accounting department, and many procedures are directly related to accounting transactions, such as payroll, accounts payable, cash, fixed assets, and purchasing. In addition, the controller's involvement ensures that appropriate checks and balances are built into the procedures.

If the procedure-writing task is assigned to the department primarily concerned, there is a danger that different departments will create manuals that do not have consistent formats. Top management should distribute a standardized format to all departments, so that at least the following items are standardized:

- Page numbering. All procedure pages must be numbered. Otherwise, users will not know if steps are missing.

- Revision dates. Procedures are revised continually. Without revision dates on every page, users may follow outdated guidelines.

- Step numbering. Long procedure steps are difficult to follow. Guidelines should have examples of short steps with opening, explanatory, and closing sentences.

- Flowcharting guidelines. Flowchart symbols may mean different things in different departments. If a rectangle represents a document in one department and a process in another department, external users might be confused.

The time needed to create policy and procedures manuals is appreciable, and it is imperative that responsibility for manuals be properly placed and that they be efficiently created to achieve maximum return on the investment.

Example Policies

As noted earlier, a policy is a statement by top management to be used as a general guide by the organization on how to deal with a situation. Figure 10.10 lists several policies pertaining to the order-taking example. The policies in the figure clearly state the boundaries within which personnel may operate. For example, requested credit of $50,001–$100,000 must be approved by the credit supervisor. An explicit statement like this is less subject to misinterpretation than "excessive credit must be reviewed by management." Clearly stated policies lead to minimal staff confusion, speedier transaction processing, and minimal errors to correct.

Policy: Credit Limits

June 6, 1996   **Page 1 of 1**

Any orders received that are not prepaid and exceed $50,000 will be forwarded to the credit supervisor for credit approval. Any such orders that exceed $100,000 will be forwarded to the Treasurer for additional approval. Any such orders that exceed $250,000 will be forwarded to the Chief Financial Officer for additional approval.

Policy: Order Filling

June 6, 1996   **Page 1 of 1**

All orders with stock on hand will be filled within two days of order receipt. If stock must be back-ordered, the order will be shipped with all possible items within two days of order receipt; back-ordered items will be shipped within one day of receipt.

Policy: Inventory Audits

June 6, 1996   **Page 1 of 1**

All on-site raw materials inventory will be audited once a week. The audit team will review the quantity, unit of measure, and location data for a statistically significant proportion of the inventory. All off-site inventory will be reviewed monthly, using the same criteria.

FIGURE 10.10   Order-taking and related policies.

Example Procedures

As noted earlier, a procedure is a specific rule or series of steps to follow in accomplishing a task. Figures 10.11–10.15 list several procedures pertaining to the order-taking example. All are very specific. Without clear instructions, an employee will have to guess at the correct sequence of events and will likely create errors by doing so. Consequently, it is better to write procedures with too much detail than with too little.

Steps in Preparing a Manual

Effective manual creation starts at the top of the organization. Management must determine which manuals should be created. Without this high-level decision, unnecessary manuals may be produced. Manuals are generally needed for all departments, but management can determine the priority in which manuals will be created.

Procedure: Order Taking

June 6, 1996   Page 1 of 1

If Payment Is by Credit Card

1. Enter name as listed on credit card.
2. Enter type of credit card.
3. Enter credit card number.
4. Enter expiration date.
5. Enter product number.
6. Verify product description with customer.
7. Enter product quantity.
8. Calculate total cost.
9. Verify cost with customer.
10. Verify ship-to address.

If Order Is by Purchase Order

1. Enter company name.
2. Enter company phone number.
3. Enter company address.
4. Enter company references.
5. Enter product number.
6. Verify product description with customer.
7. Enter product quantity.
8. Calculate total cost.
9. Verify cost with customer.
10. Verify ship-to address.

FIGURE 10.11   Order-taking procedure.

Procedure: Credit Review

June 6, 1996   Page 1 of 1

1. Contact Dun & Bradstreet automated credit check at (303) 123-4567.
2. Enter the phone number (including area code) of the company under review.
3. Ask to have a payment history sent to our fax number at (303) 765-4321.
4. If average payment days are below 40, accept credit up to a limit of $50,000.
5. If average payment days are over 40, review payment history on previous receivables. If average payment days are below 40, accept credit up to a limit of $50,000.
6. If there is no previous receivable from the company under review, give the file concerning the company to the credit supervisor for further review.

FIGURE 10.12   Credit review procedure.

Procedure: Send Order Acknowledgment

June 6, 1996   Page 1 of 1

1. Access the Order Entry software, and enter the Review Inventory Status screen.
2. Enter the part numbers and quantities from the customer order.
3. The software will access the inventory database and reserve inventory for that order.
4. If necessary, enter Y for items that must be placed on back-order.
5. Once the order has been completely entered, enter Y at the Print Order Acknowledgment? prompt.
6. The Order Acknowledgment will be printed in duplicate.
7. On the customer's copy of the Order Acknowledgment, highlight any back-order items with a yellow marker, and mail to the customer.
8. File the remaining copy of the Order Acknowledgment by customer last name.

FIGURE 10.13   Send order acknowledgment procedure.

Procedure: Ship Order

June 6, 1996   Page 1 of 1

1. Access the Inventory software, and enter the Fill Orders screen.
2. Access the oldest unfilled order (press F2 for list of orders sorted by age).
3. Print out the kitting list for that order by entering Y at the Print Kitting List? prompt.
4. The kitting list is sorted by bin location. For example, start at aisle A and pick from both sides of the aisle as you progress down the aisle. As items are picked, check off the item on the kitting list. If items are not available, circle the item on the kitting list and enter the partial amount picked.
5. Return to the shipping station. Enter the items picked in the Fill Orders screen.
6. Enter Y at the Print Packing Slip? prompt. The packing slip will be printed on the printer next to the shipping station, along with a shipping label.
7. Enclose the order in container of the appropriate size (all shipping containers are located next to the shipping station), fill excess space with shipping popcorn, and add the packing slip. Then tape the container closed, affix the shipping label to the outside of the container, and place in the UPS delivery box.

FIGURE 10.14   Ship order procedure.

Procedure: Compare Orders Received to Orders Shipped

June 6, 1996   Page 1 of 1

1. The internal auditor will procure the Orders Received log from the sales department.
2. Access the Inventory module, and enter the Fill Orders screen.
3. Either on a sample basis or in total, take a customer name from the Orders Received log and compare the initial order to the amount listed as being shipped in the Fill Orders screen for that customer.
4. If the amount shipped varies from the amount ordered, access the back-orders screen and, entering the customer name, cross-check the amount back-ordered with the variance between the amount ordered and the amount already shipped.
5. If the customer has waited more than two weeks for the back-order to be shipped, note this in the audit report. If all or a portion of the customer order has been neither shipped nor back-ordered, note this in the audit report. If incorrect quantities or items were shipped to the customer, note this in the audit report.

FIGURE 10.15   Compare orders received to orders shipped procedure.

The following general points should be considered when preparing a manual:

1. Determine who will use the manual and how it will be used.
2. Determine the manual's objectives, and its contents based on those objectives.
3. Prepare a rough outline of the manual.
4. Define problem areas for each procedure, and note alternative decision points.
5. Collect all forms and reports related to each procedure.
6. Review procedure steps currently being used.
7. Prepare a draft of each proposed procedure. Drafts should provide sufficient space for reviewers to make changes. Titles should be used in lieu of personal names.
8. Secure comments on the draft from all interested departments.
9. Prepare a revised draft reconciling conflicting viewpoints and incorporating suggestions made, to the extent practicable.
10. Establish and obtain the approvals required to publish the completed procedure. The following approvals are recommended.

Title	*Area Requiring Approval*
President or CEO	All statements of basic policies
Division general executive	All procedures implementing corporate policies relative to functions under the position's control
Controller	All procedures that relate to accounting and control, including cash, cost transfers, and the recording of all economic events
Procurement executive	All procedures concerning the purchase of materials or services
Industrial relations executive	All procedures relating to personnel, employee benefits, safety, labor relations, and compensation
Research & engineering executive	All procedures relating to technical activities

In summary, all basic policies must be approved by the CEO, those relating to specific functions by the function head, an those of an operating unit by the manager of the unit.

11. Determine the distribution list.

12. Prepare and distribute the manual.

Contents of an Accounting Manual

An accounting manual should set forth all the accounting policies and related procedures. The content and organization of such a manual will vary according to the size of the business unit, nature of the business, and organizational structure; however, the following is illustrative of a format and the subjects to be covered:

General

- Purpose and use of manual.
- Methods for making and approving changes.
- Organization charts—accounting and finance function as well as reference to the overall organization of the business unit and the company.

- Functional outlines of all accounting units.
- Job descriptions of accounting and finance positions.

Financial Policies and Procedures

- All policies related to finance and accounting as well as detailed implementation procedures.

Accounting Procedures

- Chart and text of accounts, including classification and a general description of each account.
- Detailed implementation procedures.

Accounting Closing and Reporting

- Closing procedures and schedules.
- Internal and external reports prepared, stating designated responsibility, due date, and distribution.
- Master schedule of reports required by and from the controller's organization.
- Securities and Exchange Commission reports.

Specialized Policies and Procedures

- Specialized reports such as cost determination and allowability on government contracts.
- Property control.
- Foreign exchange.
- Payroll.
- Inventory.
- Government contracts.

Accounting Documents

- Standard accounting forms such as payment authorizations, travel expense reimbursement, budget authorization, customer invoices, and intercompany invoices.
- Standard report formats including profit and loss statements and department expense reports.
- Standard letters (e.g., collection follow-ups in series, bank transfers).

Job Descriptions

Clearly, a manual must tell employees how to do their jobs. However, mistakes can also occur in the area of *who* does the work, especially in high-turnover situations where the staff is not familiar with the allocation of various tasks. Consequently, job descriptions are necessary.

A job description should list the following items.

- Update date. Tasks, staff responsibilities, and reports change constantly, so the job description must be updated frequently to reflect the alterations.
- Reports to. Surprisingly, many employees do not know who their supervisor is, or alternatively, which of many apparent bosses is the real one. Identifying the appropriate person eliminates any confusion regarding who is authorized to assign tasks to the employee.
- Staff reporting to. Once again, identifying one's staff is necessary when assigning work and also avoids conflict with other supervisors regarding multiple bosses assigning work to the same individual.
- Tasks responsible for. Tasks can be listed under subheadings, such as daily, weekly, monthly, and annual tasks. The procedure name or reference number can be listed next to each responsibility, so the person can quickly reference more detailed information about each task. Reports generated should also be listed.

Figure 10.16 is an example of a job description for a controller of a small company, with additional responsibilities in the areas of treasury, human resources, management information systems, and office management.

Distribution of the Manual

The individual or organization responsible for the manual should maintain an up-to-date distribution list of all manuals. The list should be reviewed periodically to determine that distribution is being made to those with a real need. To the extent that a manual contains proprietary or classified material, controls should be established so that the manuals are only distributed to approved personnel and are not available to outsiders. Care must be exercised so that manuals are returned to the originator when employees no longer have a need for them or when employees are terminated from the company.

Manuals can also be published electronically. The manuals can be stored in a central database, distributed by electronic mail, or stored and distributed on a

---

Title: Controller

June 6, 1996 Page 1 of 1

Reports to

- President

Staff Reporting to

- General Ledger Accountant
- Receivables Clerk
- Payables Clerk
- Payroll Clerk
- MIS Supervisor
- Receptionist

Daily Tasks

1. Review daily cash balance.
2. Ensure that daily system backup was completed (see Backup procedure).

Weekly Tasks

1. Update weekly cash forecast.
2. Review schedule of MIS projects with MIS supervisor, and alter priorities as necessary.
3. Attend executive committee meeting.

Monthly Tasks

1. Review general ledger journal entries.
2. Review financial statements.
3. Prepare financial statement commentary and issue to shareholders along with financial statements (see Financial Statement Commentary procedure and Monthly Financial Statement Report).
4. Review tax schedule for tax filings due (see Tax Calendar).
5. Compare voicemail listing of employees to payroll list, and adjust voicemail list (see Voicemail Maintenance procedure).
6. Review procedures manual for updates, and issue updates as necessary.
7. Review personnel records and notify supervisors of upcoming performance and pay reviews (see Personnel Performance Review procedure and Personnel Pay Adjustment procedure).
8. Ensure that monthly system backup was completed (see Backup procedure).

FIGURE 10.16 Controller job description.

compact disc. All the traditional steps used in developing and approving policies and procedures are followed except that there is no paper flowing between individuals. The elimination of the typing, retyping, printing, proofreading, binding, packaging, and shipping is a big cost saver and an even bigger morale boost to the personnel responsible for completing such tasks.

SUMMARY

Flowcharting and related systems documentation are needed to identify the parts of existing processes that can be streamlined and should also be used for improved processes, so that employees will know the correct methods for completing the process.

Manuals are needed to tell personnel how to complete procedures in the correct manner, thereby avoiding errors that must later be corrected. Each manual should include a clear flowchart of each process as well as written descriptions of each step and coverage of the most common process exceptions. Once the manual has been completed and properly distributed, it must be enforced with periodic compliance audits to enforce adherence to the published policies and procedures. Finally, manuals are not static—they must be changed promptly as the company's processes change, so that users will always have up-to-date information.

REFERENCES

Brown, Harry L. *Design and Maintenance of Accounting Manuals.* 2d ed. New York: Wiley, 1993.

Murtuza, Athur. *The New Accounting Manual.* New York: Wiley, 1995.

Willson, James D., Janice M. Roehl-Anderson, and Steven Bragg. *Controllership.* 5th ed. New York: Wiley, 1995.

11

EFFECTS OF CHANGE ON EMPLOYEES

This chapter differs from the previous ones because it does not deal with the mechanics of streamlining accounting transactions or with the resulting problems caused by control points, new technology, error rates, or reports. Instead, it discusses the effect on employees of the many changes advocated in this book. More than a hundred changes are listed in the previous chapters, and implementing even a fraction of them will cause significant disruption in the work of the accounting department as well as in departments that interact with the accounting group. In particular, one study shows that 84% of the labor in the accounting and finance departments is devoted to transaction processing—the precise area that this book focuses on shrinking, meaning that the majority of accounting employees will be affected by these changes. This chapter warns the reader of organizational problems caused by change, details the effects of disruption on the organization, and explains how to deal with change in a positive manner.

EFFECTS OF CHANGE ON THE ORGANIZATION

Many improvement projects never reach completion because the organization rejects them. Minor changes have a greater chance of not being rejected by the organization, since employees do not feel threatened by them, but major changes require a quite different implementation approach.

A major change probably will not be accepted by an organization for a variety of reasons. First, many employees are comfortable operating within the set of procedures that defines their jobs. By changing the procedures, their jobs are changed, which causes a considerable amount of uncertainty in their lives. They will react to any such changes by trying to reintroduce their comfortable former procedures. Second, many changes will cost employees their jobs; reaction to this type of change by the affected employees is understandable. Also, changes may require new lines of authority, so that employees find themselves reporting to new supervisors. Finally, certain types of employees react negatively to all types of change and will continue to do so no matter how well they are educated about the reasons for change. Any or all of these problems will surface during the implementation of a major change.

When a major change is in the process of implementation, the organization will pass through a well-defined set of responses. The first response is only by the management team, which is optimistic that the group is finally working toward a worthwhile objective. Then, after a detailed analysis of the project requirements, pessimism regarding the extent and cost of change sets in—this affects the entire set of employees affected by the project, not just the management team. Many projects fail at this point. Then, as the project proceeds and roadblocks are overcome, a cautious sense of optimism begins to pervade the organization, optimism that grows as the project nears completion. There will then be a period of rejection at the end of the project when the new system replaces the old system and questions are raised about various problems (perceived or real) with the new system. It is still possible for the project to fail at this point. Finally, after the last objections are overcome, the organization accepts the new system.

When an organization makes lightning-fast changes, it becomes so unstable that the work force becomes transient. When whole departments are formed and dissolved over very short time periods, it becomes evident to the employees that the company is not willing to provide any tenure on the job. Therefore, they tend to become focused on their positions—becoming the best accountants, marketing managers, engineers, and so on. This is good to an extent, because the company ends up with a highly skilled work force. However, those employees who are loyal to their professions rather than to the company will not stay long, and will take their skills and knowledge of the company's procedures with them when they leave. Change results in a short-term work force.

In summary, the project must pass through several phases when the organization is likely to reject the project, and changes may result in major staff turnover and minimal employee loyalty to the company. How does the management team ensure that projects will survive this obstacle course of emotional trauma and reach a successful conclusion? The next section offers some suggestions.

HOW TO OVERCOME ORGANIZATIONAL RESISTANCE TO CHANGE

Project implementation teams tend to focus on the technical aspects of their projects and are taken aback when the organization fails to support their projects for reasons entirely unrelated to the benefits of the projects. However, by being aware of the various reasons why organizations reject projects, it is possible to circumvent the problems presented by change.

The first step the management team should take is to ensure that a project has a strong sponsor. This must be a senior manager who is deeply committed to the project, and who is willing to intercede for the project team whenever necessary to ensure that the project will be a success. When the organization realizes that this senior manager is totally committed to the project, it will realize that this project is "inevitable" and "cave in" to the desires of the project sponsor. However, this can be construed as forcing the project onto the organization. It is better also to persuade the organization to accept the changes. How can this be done?

The project team must meet with a variety of employees who will be affected by their new project and show them the problems that will occur if the project is *not* completed on time, such as higher costs that lead to future layoffs or longer processing times that make the company less competitive and that also may lead to layoffs. The delineated crisis must be real or else employees will suspect that management is making up reasons for forcing the project upon the company. This gloomy view of the future can be followed by a clear picture of the benefits to be gained from the new system as well as by a detailed, step-by-step plan for how to get to that point. This information should be laid out in as much detail as possible, so that employees know that the project team (and management) is not hiding any information from them. As the project progresses, members of the project team should meet with the company's employees from time to time and update them on the project's progress. This sharing of information helps to gain the acceptance of the organization for the project.

If information sessions are not enough, the project team should schedule education sessions for key users from the organization that gives them an in-depth view of the changes being implemented. Training can come from many sources. The information contained in training classes can come from experimenting with new ideas, comparisons with other (possibly unrelated) organizations, and formal training courses. This level of involvement is sufficient to persuade the majority of employees that the project should be implemented.

There will always be a few employees who are unchangeably comfortable with the status quo and who will attack any proposed changes that will alter the current set of procedures that defines their jobs. The project team should listen to the concerns of these people and address any problems that can be accommodated without jeopardizing the successful (and timely) completion of the project. However, if these people continue to oppose the project after all reasonable efforts have been made to accommodate them, they should be removed from the part of the company that is affected by the project. This is a painful step to take, but is necessary to avoid continuing sniping at the project by those employees even after the project has been implemented.

If some employees are shifted out of the project area as a result of a new project, new employees must be hired into the focus area. This is a considerable opportunity for the project team and company management to hire the right kind of employees, who readily accept change and who are comfortable in that type of environment. Interviews for these new positions should be rigorous, focusing in particular on the candidates' level of comfort in a changing environment, current technical skills, and willingness to assist further change efforts.

Many new proposals will result in a reduction of the company work force. Rather than leave employees guessing about who will be asked to leave the company, it is more fair to define standards of performance for the company staff as soon as possible that define how the company will choose who stays when a layoff becomes necessary. This performance measurement system allows employees to measure themselves against the standard and determine for themselves if they will be laid off. This system greatly reduces the uncertainty of the staff during a period of intense competitive pressures and ensures that management can rank its employees so that the most productive are retained as the core group of necessary employees.

Finally, fully implemented projects may still fail if the compensation system is not changed to support it. If a new system is not supported by an altered compensation system, then employee behavior may be so constrained by the compensation system that it is impossible for the changes to be accepted by the organization. Involving the human resources and management team in this aspect of the project implementation will ensure that supportive compensation plans are in place once the project has been completed.

If the project team has a strong sponsor, communicates a detailed project plan to the organization, removes "snipers" from the focus area, hires new employees who are comfortable with change, and alters the compensation system to support the changes, then the project team has a good chance of success.

THE CHANGE MANAGER

This book is about implementing changes, not rising to the top of the corporate hierarchy. As process improvements take hold in corporations, those with an ability to implement the changes will be more valuable than those who supervise continuing tasks. Currently, a higher pay level is awarded to the manager who controls a department with lots of staff and expensive fixed assets such as mainframe computers, whereas lower pay goes to those employees who run lean operations with minimal staffing and fixed assets. If this compensation system continues, managers may be concerned about improving the efficiency of their departments, since the resulting reduction in assets and personnel may lead to a "demotion" in pay.

Since a more efficient, flattened organization has fewer managers, people in the work force will have a more difficult time achieving exceptional compensation by rising through the corporate hierarchy—after all, there won't be much of a hierarchy to rise through. Instead, these employees must derive their satisfaction based on projects completed, skills gained (both on the job and through the traditional methods of earning college degrees and other certifications), and problems overcome.

In companies with many layers of management, most managers spent their time (and are rewarded for) communicating within their areas of specialization; for example, the accounting department is only concerned with having perfect accounting controls, while the production department tries to meet its production volume controls. In a flatter hierarchy, it is logical that managers must now communicate sideways (between departments). This means that problems caused by the self-centered behavior of individual departments are exposed by the "new" departments with whom they are now communicating. For example, the production department in a flat organization will realize by talking to the sales department that it is meeting production volume goals but only at the price of shipping defective products, which hurts the sales of the company when netted against returns and the long-term prospect of customer ill-will.

In the new, rapidly changing world of the highly efficient corporation, many companies will relegate transaction-processing management tasks to low-grade managers and promote their star managers into project management roles. These are projects with definite beginnings and endings, with specific deliverables. A project manager will not compile a portfolio of previous job titles but rather a portfolio of completed projects. As more change projects appear in an organization, the project manager role proliferates. This is a difficult position, for the project manager must handle a team drawn from multiple departments, achieve the goal with a limited budget, and will be out of work when the project is completed, hopefully going on to an even larger project with more visibility (and risk) that will lead to ever-larger projects. The unique skill of the project manager is in project management and in handling a diverse set of department managers who may be funding the project and who are likely to be contributing their staff to the effort—these people must be kept happy so that resources will continue to flow into the project, resulting in a successfully attained target.

A unique variation on the change manager is not an individual but a group of top-notch employees, the "hot group." The hot group is a small group of employees who are totally dedicated to a particular task. There is not necessarily a leader—this role may change as the task progresses. The hot group will organize itself and prioritize tasks without outside assistance. These groups are usually short-lived, terminating when their task is complete. They originate in companies that do not have stifling administrations in place. Team members come from multiple departments—this means that hot groups can only form in companies where different departments are allowed to intermingle. Hot groups should not be supervised closely, which is hard for formal managers who want regular status reports and meetings. Hot groups form most readily in crisis situations where the normal plodding form of interdepartmental decision making is the norm. The people who form these groups are typically possessed of strong egos and don't mind sharing the credit for work performed. The leaders do need some skills in managing groups, however—as is the case with any management role, it is important to involve all team members in the decision-making process and to acknowledge their input in the final decision. This leads to better group commitment to the final decision, attachment to the group, and trust in the group leader. Organizations do not like hot groups because they work across boundaries at high speed, flouting normal conventions, such as reports and feedback sessions; they also work at odd hours and give the impression of being prima donnas. Given the problems with forming and maintaining hot groups, it is clear that the top management group must be highly supportive of these unique creatures of the corporate environment in order to see worthwhile results from their efforts.

In short, the traditional manager will be relegated to the corporate junkyard of managing transaction-based operations while the top-notch managers will lead project teams that push change through the organization. A unique variation on the top-notch manager will be the hot group, which has no clear leader but is capable of generating top-quality results by being a dedicated group of employees who bring projects to a rapid and successful conclusion.

ULTIMATE CHANGE: THE REENGINEERING PROJECT

A reengineering project usually involves a radical redesign of an existing process. Radical change requires considerable changes to the organization, which can cause considerable disruption among employees. This section discusses how to implement a project that involves a great amount of change.

Reengineering projects fail more frequently than they succeed. Part of the problem is that reengineering is not an incremental change—it involves tearing out the old system by the roots and installing an entirely new system that may not mesh very well with the existing organizational structure. To make a reengineering project more successful, it is necessary to convince as many employees as possible of the need for change. Arguments can include loss of market share, spiraling costs, and a poor showing in comparison to the industry benchmark. Even a tour of a company that is world-class in regard to that process may be needed to convince unbelievers.

Some of the people requiring convincing will be top management. If the management team does not throw its support behind the project, as well as assign one of its members direct responsibility for it, the project should not be undertaken. This is because top management may pull out all monetary and personnel support of the project if it can be convinced that other projects are more important or that the benefits to be derived from the project will not occur. Instead, the top managers must be so enamored of the project that they are constantly backing it when meeting with other members of management, inquiring about it, and pushing for earlier completion dates.

A problem encountered when convincing managers of the need to change is their level of discomfort with the degree of change involved. They may be willing to make a few alterations to procedures, but the wholesale dismantling and reassembling frequently encountered in a reengineering project may be more than they are willing to support. One way to avoid this problem is to reengineer on a pilot basis, so that the company can see how the reengineering project is affecting the performance of a small area of the company. If the pilot project goes well, then acceptance of the project as a whole is much more likely. If the pilot project does not proceed according to plan, then it may not have worked on a large scale anyway and should be refined at the pilot project level before being expanded throughout the organization.

A more acceptable approach for the existing organization is to come up with radical changes at the planning stage but to implement the changes slowly and in segments, so that the organization has a better chance to assimilate the changes. However, this slow-change model embodies the risk of being too slow. If the original project is watered down and strung out over a long time, the project team supporting the project may become discouraged and drift off to other projects. Also, if a company is in dire need of change, the implementation must proceed at full speed in order to keep the company from failing—the effect of the changes on the organization become a secondary consideration.

Another problem with reengineering is its dependence on advanced technology to add efficiency to a process. If the technology is too advanced, then it may not stand the test of actual usage on a day-to-day basis. To avoid this problem, the company should conduct pilot tests of new technology or at least send a team to a company that has successfully implemented the technology to inquire into any problems that may have been encountered during implementation. Another solution is to back away slightly from the most advanced technology and instead adopt technology that has been tested in the marketplace for a few years.

A reengineering project may fail if it is not staffed with an adequate number of people who are well trained and skilled in completing the project. This is particularly important for a reengineering project where advanced technology is frequently used, skills for handling recalcitrant employees are paramount, and project management skills for handling a diverse group of highly talented individuals is necessary. Project management skills are especially hard to come by. A top project manager must be able to keep the project within a time and a money budget, schedule and administer project reviews, coordinate subprojects, and communicate about the project's progress with upper management. Keeping the expectations of the top management group in line with reality is particularly important, since there should be no fallout at the end of the project if the system delivered to the company does not meet with its expectations. At lower organizational levels, the project manager should also be involved with obtaining acceptance of the project by those individuals or departments that will be affected by the reengineering project. If there is resistance in those departments, the successful conclusion of the project will be in jeopardy.

Perhaps the most important issue for a reengineering project is that it must be planned adequately to convince the many nay-sayers who will arise when the project is announced. The plan must detail the project's staffing needs, time line, role-out schedule, training requirements, costs and benefits, and (especially) effects on the existing staff. It is best to be as detailed and forthright with this information as possible, both to convey an image of integrity to the staff (even those whose jobs may disappear as a result of the changes) and to provide information to employees who may offer valuable advice that may result in changes to the plan and contribute to its ultimate success.

In short, the reengineering project, which is the most difficult project of all to implement, can be completed by following the guidelines listed in the section that discussed overcoming organizational resistance to change. The main issue that sets apart a reengineering project from other projects is that *all* the guidelines must be followed, for the organization most certainly will not accept the project otherwise.

ROLE OF TOP MANAGEMENT IN IMPLEMENTING CHANGE

A company's top management team will find that its methods of controlling the business will not help the organization to absorb rapid change, which means that top management's way of doing business must change to accommodate the company's conversion to a change-based organization.

Top management is used to tightly controlling the direction of the company by requiring lower management to report results in great detail, both through computerized reporting systems and formal structured meetings. The willingness of lower managers to present innovative new approaches for running the business is reduced when they are preoccupied with providing reasons for variances that are generated by the formal reporting system as well as with preparation time for formal meetings that strictly follow predetermined agendas. In short, the formal control system tends to inhibit creativity and initiative.

An alternative approach for senior management is to avoid the formal meeting and instead to have frequent personal contact with their subordinates, which provides a forum for the junior managers to make recommendations that might otherwise never be presented to senior management. Also, the information-reporting system can be turned around and presented to the lower levels of management for their decision making. This frequently forces the lower levels of management to arrive at the same conclusions that senior management arrived at with the same information, which engenders unity of purpose across the management hierarchy of the company.

Also, as the senior management group learns to stop controlling lower management so tightly, it finds that there is an alternative way to encourage the employees of the company—by increasing their time devoted to human resources topics. This shift of focus requires senior management to think about where to assign the company's brightest employees. If the company's strategic direction lies in a certain subsidiary of the company, senior management should ensure that the company's best people end up in that subsidiary—this is a better approach than simply theorizing about strategic direction in the corporate headquarters. Another human resources issue for the senior management team is the reward structure used to motivate the lower management group. By subtly altering the reward system, senior management can have a direct effect on the actions of company personnel.

Finally, senior management must avoid reliance on information systems, since they only report on information that has already occurred and tell senior management absolutely nothing about what to except in the future. For the best clues about that, it is still best for senior management to network among the lower levels of management, which may have a better grasp of what is going on in the market.

In short, the top management group must reduce its reliance on the formal reporting system and instead focus on attaching the company's best employees to the most important projects as well as pushing information back down to lower-level managers, who can then make decisions based on better information.

EFFECTS OF STRETCH GOALS ON THE ORGANIZATION

Many firms are now using stretch goals as a means of motivating employees into making great leaps forward in productivity. A stretch goal is a target that is so difficult to reach that entirely new means of conducting operations must be invented to reach the goal. Though a successfully completed stretch goal can energize a company and rapidly take it to highly efficient levels of operations, it can also have several less welcome effects on employees. First, if a stretch goal is imposed without any accompanying support in the form of funding, the staff may look upon the goal as a veiled message to leave the firm. This message may appear to be confirmed if the company imposes strict penalties on failure. If employees do reach the goal with such support, it is probably because they are spending more of their personal time (the only resource they have that is free and available in some quantity) on the work; this results in overworked and stressed employees. Also, it is not a good idea to impose stretch goals on the company's most productive employees because the most efficient employees have already figured out how to perform tasks in their assigned areas with an economy of motion; imposing a stretch goal on them will almost certainly not result in the goal being achieved while imposing great stress on them. Also, since the high achievers (who are used to completing their goals) will not have achieved the stretch goals, they will now feel like losers. With a negative effect on morale, the overall performance of the organization may decline.

In summary, stretch goals must be carefully developed and targeted at the areas of the company that have the most "fat," so that the prospects of attaining the goals are not completely impossible. Also, the company must devote sufficient resources to the effort to make the odds of success fairly high while not penalizing employees if they put forward their best efforts but do not achieve their goals.

SUMMARY

As more companies attempt to bring change to their employees, they find that many projects are not being successfully completed. This low success rate is caused by the discomfort the organization experiences when existing systems are replaced by new procedures, technologies, and reporting relationships. These problems can be overcome by gaining strong management support, assigning the best employees to project teams, planning each project in detail, communicating change plans to employees, moving dissenters away from the focus areas, and altering the compensation system to support the new systems. These changes improve the odds of a project's success.

REFERENCES

Bartlett, Christopher A., and Sumantra Ghoshal. "Changing the Role of Top Management: Beyond Systems to People." *Harvard Business Review,* May/June 1995, pp. 132–142.

Birkett, W. P. "Management Accounting and Knowledge Management." *Management Accounting,* November 1995, pp. 44–48.

Bloch-Flynn, Pamela, and Kenneth Vlach. "Employee Awareness Paves the Way for Quality." *HR Magazine,* July 1994.

Boyle, Robert D. "Avoiding Common Pitfalls of Reengineering." *Management Accounting,* October 1995, pp. 24–33.

Bridges, William. *Surviving Corporate Transition.* Mill Valley, Calif.: William Bridges & Associates, 1993.

Building Commitment to Organizational Change. Denver: Ernst & Young LLP, and ODR, 1992.

Conner, Daryl R. *Managing at the Speed of Change.* New York: Villard Books, 1993.

Determinants of Successful Organizational Change. Denver: Ernst & Young LLP, and ODR, 1992.

Dumaine, Brian. "Times Are Good? Create a Crisis." *Fortune,* June 28, 1993.

The Emotional Cycle of Change. Denver: Ernst & Young LLP, and ODR, 1992.

Ernst & Young U.S. LLP Performance Improvement Series: Organizational Change Management Methodology. Denver: Ernst & Young LLP, 1992.

Hackett Group. *AICPA/THG Benchmark Study: Results Update and Analysis.* 1994.

Hammer, Michael, and Steven Stanton. "Beating the Risks of Reengineering." *Fortune,* May 15, 1995, pp. 105–114.

Hefner, Mark, and Dave Schrader. "Merging Effectively: Taking the Mystery Out of Managing Change." *Current Issues,* June 1994.

How to be an Effective Sponsor of Organizational Change. Denver: Ernst & Young LLP, and ODR, 1992.

Implementation History Assessment. Denver: Ernst & Young LLP, and ODR, 1992.

Keily, Thomas J. "Why Reengineering Projects Fail." *Harvard Business Review,* March/April 1995, p. 15.

————. "Reengineering: It Doesn't Have to Be All or Nothing." *Harvard Business Review,* November/December 1995, pp. 16–17.

Kotter, John. "Leading Change: Why Transformation Efforts Change." *Harvard Business Review,* March/April 1995, pp. 59–67.

Leavitt, Harold J., and Jean Lipman-Blumen. "Hot Groups." *Harvard Business Review,* July/August 1995, pp. 109–116.

Merlyn, Vaughan. *When Change Becomes Continuous.* Denver: Ernst & Young LLP, 1994.

Oakley, Ed, and Doug Krug. *Enlightened Leadership.* Denver: StoneTree Publishing, 1992.

Penzias, Arno. "New Paths to Success." *Fortune,* June 12, 1995, pp. 90–94.

Predicting the Impact of Change. Denver: Ernst & Young LLP, 1993.

Rheem, Helen. "Building Learning Capability." *Harvard Business Review,* March/April 1995, p. 10.

Richman, Tim. "Strategic Decision Making: It's All in the Process." *Harvard Business Review,* September/October 1995, pp. 10–12.

Role Map Application Tool. Denver: Ernst & Young LLP, and ODR, 1992.

Ruchala, Linda V. "New, Improved, or Reengineered?" *Management Accounting,* December 1995, pp. 37–42.

Sherman, Strat. "Stretch Goals: The Dark Side of Asking for Miracles." *Fortune,* November 13, 1995, pp. 231–232.

Stevenson, Howard H., and Mihnea C. Moldoveanu. "The Power of Predictability." *Harvard Business Review,* July/August 1995, pp. 140–143.

Stewart, Thomas A. "The Corporate Jungle Spawns a New Species: The Project Manager." *Fortune,* July 10, 1995, pp. 179–180.

X-Factor Change Readiness Assessment Technique. Denver: Ernst & Young LLP, 1994.

APPENDIX

SUMMARY OF RECOMMENDATIONS

Many recommendations are scattered throughout this book. In this appendix, the key streamlining suggestions from Chapters 1–9 are listed to facilitate review by the reader. Those suggestions considered to be common knowledge (e.g., tracking receivables with the days of receivables measure) are not included here. No recommendations are listed for the final two chapters, since those chapters are devoted to describing the flowcharting existing systems (Process Documentation) and anticipating the organizational problems that will occur when changes are implemented (Effects of Change on Employees).

Chapter 1—Cash

- Implement a lockbox system.
- Implement an area concentration banking system.
- Start a zero balance account.
- Eliminate compensating balance agreements with banks.
- Implement an electronic data interchange (EDI) system.
- Reduce the number of approvals in the purchasing and payables system.
- Pay for more items through the petty cash system.
- Install a computerized accounting system.
- Enforce a minimum dollar limit for approval of purchases or accounts payable.
- Track invoice payment dates with a computer database.
- Have the computer software store the next check number and print it on the check.
- Automate the application of cash to receivables with an EDI interface.
- Have the computer software calculate invoice payment due dates.

- Have the computer software match supplier invoices to purchase orders for payment.
- Pay from purchase orders instead of supplier invoices.
- Send invoices using EDI transactions.
- Have the computer software spot duplicate supplier invoices.
- Report on customers and suppliers with large order volumes (for EDI conversion).
- Compare the check date to the clearing date to track float time.

Chapter 2—Sales and Accounts Receivable

- Move transactions between employees by electronic mail.
- Preapprove customer credit.
- Hand the invoice to the customer when the product is delivered.
- Eliminate the mailing of month-end statements.
- Immediately update receivables records when they are received.
- Consolidate the number of receivables and collection systems.
- Automate the collections tickler file.
- Send invoices to customers using EDI transactions.
- Call customers about invoices prior to the invoice due dates.
- Have customers pay based on approved purchase orders.
- Report on orders in the computer software with no related invoice.
- Report on preapproved customer credit.
- Report on received EDI acknowledgments for invoices.
- Measure the time from receipt of order to production scheduling.
- Measure the time from receipt of cash to updating of receivables records.
- Measure the number of value-added steps in the transaction cycle as a percentage of the total number of steps.

Chapter 3—Inventory

- Combine the quantity and actual cost databases.
- Move fasteners to the production area.
- Pay suppliers based on production records.
- Require a purchase order for incoming items.
- Eliminate obsolete inventory.
- Use just-in-time (JIT) manufacturing techniques.

- Use accurate bills of materials to cost products.
- Maintain highly accurate perpetual inventory records.
- Create a segregated area for consigned inventory.
- Certify suppliers.
- Limit access to unit of measure information.
- Provide multiple units of measure for inventory parts.
- Use bar codes for receiving data entry.
- Use computer limit checks.
- Limit access to bill of materials records.
- Create overlapping bill of materials change sources.
- Measure the cost of maintaining the inventory.

Chapter 4—Accounts Payable

- Simplify the ordering process for commonly purchased items.
- Automate the expense report processing function.
- Pay suppliers from the purchase order instead of the invoice.
- Eliminate approvals for purchase requisitions.
- Pay the supplier based on production records.
- Pay suppliers immediately.
- Shrink the number of accounts payable systems.
- Use corporate purchasing cards.
- Report on the payment dates of long-term contracts.
- Create office and manufacturing supplies order forms.
- Report on pricing and parts quantity disputes with suppliers.
- Report on missing expense report receipts.
- Report on transaction times for electronic expense reports.
- Measure the percentage of payments with related purchase orders.
- Measure the percentage of total payments made from production records.
- Measure the speed of expense report turn-around.
- Measure the number of payables transactions per payables employee.

Chapter 5—Cost Accounting

- Review targeted materials costs versus actual costs.
- Review the average labor rate.

- Review the amount of overhead capitalized.
- Move activities into the period prior to the month-end closing.
- Create a volume variance table.
- Revise the monthly variance report.
- Measure inventory accuracy.
- Measure bill of materials accuracy.
- Measure labor routing accuracy.
- Measure the variation of actual costs incurred to targeted costs incurred.

Chapter 6—Payroll

- Give employees direct access to their deduction records.
- Create a policy of not allowing company purchases of merchandise for employees.
- Adopt a simpler commission calculation system.
- Install a bar-coded time card system.
- Consolidate the payroll, benefits, and human resources databases.
- Reduce the number of payroll systems.
- Use direct deposit to pay employees.
- Send payroll payment advices to employees via electronic mail.
- Include the vacation and sick time accrual on the payroll remittance advice.
- List the employees who are not using direct deposit.
- List the manual changes to payroll records made by the payroll department.
- Measure the volume of data being manually input to the payroll system.

Chapter 7—The Budget

- Limit the budget range with the maximum amount of available funding.
- Do not budget detailed expense items until the high-level budget has been approved.
- Include an analysis of working capital in the high-level budget.
- Break out all fixed costs.
- Create and distribute a budget manual.
- Determine how well budgeted projects match the company's strategic goals.

- Reduce the budget from a monthly to a quarterly plan.
- Reduce the number of accounts listed in the budget.
- Use a flexible budget based on changes in the revenue level.
- Determine the drivers of key costs.
- Link a performance measurement and reward system to the budget.
- Summarize the line items of small-dollar department budgets.
- Change the color of user-defined cells in the budget spreadsheet.
- Cluster all key variables at the front of the budget model.
- Use video conferencing to reduce the number of single-location meetings.
- Measure the total number of hours spent to create the budget.
- Measure the number of accounts listed in the budget.

Chapter 8—Electronic Data Interchange

- Validate incoming EDI transactions.
- Eliminate management review of outgoing EDI transactions.
- Link EDI transactions to existing accounting systems so that manual rekeying is eliminated.

Chapter 9—The Quick Close

- Eliminate items requiring multiple approvals.
- Eliminate the filing of multiple copies of a single document.
- Flowchart the existing process.
- Eliminate duplicate activities.
- Automate recurring journal entries.
- Defer routine work.
- Set variance investigation levels that ignore small variances.
- Move closing activities into the previous month.
- Automate manual processes.
- Replace serial activities with parallel activities.
- Rearrange the work area for reduced staff move time.
- Create a closing procedure.

INDEX